THE DISCRETE SERIES OF GL_n OVER A FINITE FIELD

BY

GEORGE LUSZTIG

ANNALS OF MATHEMATICS STUDIES
PRINCETON UNIVERSITY PRESS

Annals of Mathematics Studies

Number 81

THE DISCRETE SERIES OF GL_n OVER A FINITE FIELD

BY

GEORGE LUSZTIG

PRINCETON UNIVERSITY PRESS

AND

UNIVERSITY OF TOKYO PRESS

PRINCETON, NEW JERSEY

1974

Published in Japan exclusively by
University of Tokyo Press;
In other parts of the world by
Princeton University Press

Printed in the United States of America
by Princeton University Press, Princeton, New Jersey

Library of Congress Cataloging in Publication data
will be found on the last printed page of this book

TABLE OF CONTENTS

The Discrete Series of GL_n Over A Finite Field

THE DISCRETE SERIES OF GL_n OVER A FINITE FIELD

George Lusztig

INTRODUCTION

Since the fundamental work of Green [5] it has become clear that the central role in the ordinary representation theory of the general linear group $GL_n(F_q)$ over a finite field, is played by the discrete series representations. In this work we give an explicit construction of one distinguished member, $D(V)$, of the discrete series of $GL_n(F_q)$ (here V is the n-dimensional F-vector space on which $GL_n(F_q)$ acts). This is a p-adic representation; more precisely $D(V)$ is a free module of rank $(q-1)(q^2-1)\cdots(q^{n-1}-1)$ over the ring of Witt vectors $\underset{\cdot}{W}_F$ of F.

To construct $D(V)$, we consider the top-homology $\dot{\Delta}(V)$ of a simplicial complex X associated to V; X is made out of affine flags in V which are away from 0. It turns out that $D(V)$ is naturally a direct summand (over W_F) in $\dot{\Delta}(V)$. In fact $D(V)$ is defined as an eigenspace of a certain endomorphism $T : \dot{\Delta}(V) \to \dot{\Delta}(V)$ defined geometrically, with an additional homogeneity condition. The reduction modulo p of $D(V)$ is a modular representation of $GL(V)$ which can be described as the top homology of the Tits complex of V with values in a certain non-constant coefficient system. This can be used to determine the character of $D(V)$ on semisimple elements of $GL(V)$. To deal with the non-semisimple elements we show that the restriction of $D(V)$ to any proper parabolic subgroup splits naturally in a tensor product of a module $D(V')$, $\dim V' < \dim V$, with a certain representation space of the parabolic subgroup, defined homologically using a modification of the Tits complex.

In particular, the restriction of $D(V)$ to the affine group $\mathrm{Aff}_{n-1}(F_q) \subset GL_n(F_q)$ can be described in a purely homological way, and provides a distinguished representation Δ of the affine group. This homological description of Δ has been found independently by L. Solomon ([11], [12]). The representation Δ of the affine group has been first introduced by S. I. Gelfand [4] as the representation induced by a one-dimensional representation in general position of the Sylow p-subgroup (see also I. M. Gelfand and M. I. Graev [3]).

The content of this work is roughly as follows. In Chapter 1 we study the homology of partially ordered sets and prove some vanishing theorems for the homology of some partially ordered sets associated to geometric structures. In Chapter 2 we study the representation Δ of the affine group over a finite field. In Chapter 3 we define D(V) and determine its restriction to parabolic subgroups. In Chapter 4 we compute the character of D(V) and show how to obtain other members of the discrete series by applying Galois automorphisms to D(V).

The main applications are in Chapter 5. Here we construct explicitly a Brauer lifting homomorphism $br: R_F(G) \to R_{W_F}(G)$ where $R_F(G)$ (resp. $R_{W_F}(G)$) is the Grothendieck group of representations of a finite group G over F (resp. W_F). This more refined Brauer lifting has as a simple consequence a theorem of Swan asserting the isomorphism $R_{W_F}(G) \approx R_{Q_F}(G)$ where Q_F is the quotient field of W_F (see 5.5).

Some of the results of this work have been announced in [8].

The following conventions will be used in this paper: A will always denote a commutative ring with 1 and any A-module is assumed to be finitely generated. F will always denote a finite field with q elements such that $q = p^e$. (p = prime.) Given two affine subspaces E', E'' of an affine space E, we say that E' and E'' are parallel ($E' \| E''$) if $E' \cap E'' = \emptyset$ and some translate of E' (or E'') is contained in E'' (or E'). In a sum $\sum^{(N)}$, N denotes the number of terms in the sum. The symbol V_i (resp. E_i) will always denote a vector space (resp. an affine space) of dimension i.

Finally, I would like to thank R. Carter, M. Kervaire and T. A. Springer for interesting comments.

CHAPTER 1

PARTIALLY ORDERED SETS AND HOMOLOGY

1.1 *Coefficient systems*

Let X be a finite simplicial complex. We recall the following:

DEFINITION. *A coefficient system* $\mathcal{S}$ *over* X *is a collection* $(\mathcal{S}_\sigma, \phi^\sigma_{\sigma'})$ *of* A*-modules* I_σ, *one for each simplex* σ *in* X *and of* A*-linear maps* $\phi^\sigma_{\sigma'} : \mathcal{S}_\sigma \to \mathcal{S}_{\sigma'}$, *defined for each pair of simplices* (σ, σ') *such that* σ' *is a face of* σ. *It is assumed that* $\phi^\sigma_\sigma = 1$ *and* $\phi^{\sigma'}_{\sigma''}\phi^\sigma_{\sigma'} = \phi^\sigma_{\sigma''}$ *whenever* σ' *is a face of* σ *and* σ'' *a face of* σ'.

The homomorphisms $\phi^\sigma_{\sigma'}$ will be called the *connecting homomorphisms* of $\mathcal{S}$. Let $H_i(X; \mathcal{S})$, $(i \geq 0)$, denote the homology of X with values in $\mathcal{S}$. More generally, for any subcomplex $X' \subset X$ we denote the relative homology groups by $H_i(X, X'; \mathcal{S})$, $(i \geq 0)$; these are naturally A-modules.

The following lemma can be proved easily, by a spectral sequence argument (the details will be omitted):

LEMMA. *Let* $\emptyset \subset X_0 \subset X_1 \subset \cdots \subset X_{M-1} \subset X_M = X$ $(M \geq 1)$ *be a sequence of subcomplexes of* X *such that* $H_i(X_j; \mathcal{S}) = 0$ *for all pairs* (i, j) *such that* $0 < i < j \leq M$ *and* $H_0(X_j; \mathcal{S}) \to H_0(X; \mathcal{S})$ *is surjective for* $j = 0$ *and an isomorphism for* $0 < j \leq M$. *Then there is a natural exact sequence of* A*-modules*

$$0 \to H_M(X; \mathcal{S}) \to H_M(X; X_{m-1}; \mathcal{S}) \to H_{M-1}(X_{M-1}, X_{M-2}; \mathcal{S}) \to \cdots \to$$
$$H_1(X_1, X_0; \mathcal{S}) \to H_0(X_0; \mathcal{S}) \to H_0(X; \mathcal{S}) \to 0 \ .$$

Here we denote by the same letter a coefficient system on X and its restriction to a subcomplex of X.

1.2 *Acyclic covering lemma*

Let $(X_k)_{k \in K}$ be a family of subcomplexes of X (K a finite set) such that $\bigcup_{k \in K} X_k = X$. Let N be the nerve of this covering, and let τ be a simplex of N. By definition, τ is a subset $(k_0, k_1, \cdots, k_h)$ of K such that $X_\tau = X_{k_0} \cap X_{k_1} \cap \cdots \cap X_{k_h} \neq \emptyset$. Let $N_i(\mathcal{S})_\tau = H_i(X_\tau; \mathcal{S})$. If τ' is a face of τ in N, there is clearly a natural map $N_i(\phi)^\tau_{\tau'}: N_i(\mathcal{S})_\tau \to N_i(\mathcal{S})_{\tau'}$. Then $N_i(\mathcal{S}) = (N_i(\mathcal{S})_\tau, N_i(\phi)^\tau_{\tau'})$ is a coefficient system over N. It is well known that in this situation there is a spectral sequence of A-modules which starts with $E^2_{s,t} = H_s(N; N_t(\mathcal{S}))$ and converges to $H_*(X; \mathcal{S})$. If we assume that $N_t(\mathcal{S}) = 0$ $(0 < t < a)$ where a is some fixed integer, it follows that we must have $H_t(X; \mathcal{S}) \cong H_t(N; N_0(\mathcal{S}))$ $(0 \leq t < a)$. This is the well known acyclic covering lemma.

1.3 *Direct systems*

Let S be a finite set with a partial order denoted $\leq$. S defines a finite simplicial complex $X(S)$ as follows. The vertices of $X(S)$ are the elements of S. The k-simplices of $X(S)$ are precisely the totally ordered subsets of S consisting of $(k+1)$ elements.

Let $\mathfrak{g} = (\mathfrak{g}_s, f^s_{s'})$ be a direct system of A-modules with S as set of indices. In other words, for any $s \in S$ we have an A-module $\mathfrak{g}_s$ and for any $s, s' \in S$ such that $s \leq s'$ we have given an A-linear map $f^s_{s'}: \mathfrak{g}_s \to \mathfrak{g}_{s'}$; we must have $f^s_s = 1$, $f^{s'}_{s''} \circ f^s_{s'} = f^s_{s''}$ if $s \leq s' \leq s''$. The direct system $\mathfrak{g}$ defines canonically a coefficient system $X(\mathfrak{g})$ over $X(S)$ as follows. Given a simplex $\sigma = (s_0 < s_1 < \cdots < s_k)$ in $X(S)$ define $X(\mathfrak{g})_\sigma = \mathfrak{g}_{s_0}$. If σ' is a face of σ we must have $\sigma' = (s_{i_0} < s_{i_1} < \cdots < s_{i_\ell})$ where $0 \leq i_0 \leq k$. The connecting homomorphisms $X(\mathfrak{g})_\sigma = \mathfrak{g}_{s_0} \to X(\mathfrak{g})_{\sigma'} = \mathfrak{g}_{s_{i_0}}$ are just $f^{s_0}_{s_{i_0}}: \mathfrak{g}_{s_0} \to \mathfrak{g}_{s_{i_0}}$ (note that $s_0 \leq s_{i_0}$). Define homology groups $H_i(S; \mathfrak{g}) = H_i(X(S); X(\mathfrak{g}))$, $(i \geq 0)$. Note that $H_0(S; \mathfrak{g}) = \varinjlim_{s \in S} \mathfrak{g}_s$.

More generally if S' is any subset of S with the induced partial order, we have $X(S') \subset X(S)$ and we define relative homology groups: $H_i(S, S'; \mathcal{g}) = H_i(X(S), X(S'); X(\mathcal{g}))$, $(i \geq 0)$.

DEFINITION. *Given* $s \in S$, *the weight of* s *is the greatest integer* $k \geq 0$ *such that there exist* $s_1, s_2, \cdots, s_k \in S$ *with* $s_k < s_{k-1} < \cdots < s_2 < s_1 < s$.

The weight of S is defined as the maximum of the weights of its elements. In particular the minimal elements of S are precisely the elements of weight 0. Let $S_j (j \geq 0)$ be the set of all elements of S, of weight $\leq j$. We have $\emptyset \subset S_0 \subset S_1 \subset \cdots \subset S_M = S$. ($M$ = weight of S.)

The following lemma is just a special case of Lemma 1.1:

LEMMA. *Assume that the weight of* S *is* $M \geq 1$. *Furthermore, assume that* $H_i(S_j; \mathcal{g}) = 0$ *for all pairs* (i, j) *such that* $0 < i < j \leq M$ *and* $H_0(S_j; \mathcal{g}) \to H_0(S; \mathcal{g})$ *is an isomorphism for* $0 < j \leq M$. *Then there is a natural exact sequence of* A-*modules*:

$$0 \to H_M(S; \mathcal{g}) \to H_M(S, S_{M-1}; \mathcal{g}) \to H_{M-1}(S_{M-1}, S_{M-2}; \mathcal{g}) \to \cdots \to$$
$$H_1(S_1, S_0; \mathcal{g}) \to H_0(S_0; \mathcal{g}) \to H_0(S; \mathcal{g}) \to 0 \; .$$

Note that $H_0(S_0; \mathcal{g}) \to H_0(S; \mathcal{g})$ is always surjective.

1.4 LEMMA. *Let* $\mathcal{g}$ *be a direct system with set of indices* S. *If* S *has a unique maximal element* s, *then* $H_i(S, \mathcal{g}) = 0$ $(i > 0)$ *and* $H_0(S, \mathcal{g}) = \mathcal{g}_s$.

This follows easily using a standard homotopy argument; note that $X(S)$ is contractible.

1.5 We define now a canonical covering of $X(S)$, where S is any finite partially ordered set. Let S^{max} be the subset of S consisting of all maximal elements. Given $s \in S$ define $S^s = \{s' \in S \mid s' \leq s\}$ and let $X^s = X(S^s)$. The complexes X^s $(s \in S^{max})$ cover $X(S)$. This covering will be called the maximal covering of $X(S)$.

Assume that S satisfies the following property:

(1) *Let* $s_0, s_1, \cdots, s_k \in S$. *If the set* $\{s \in S | s \leq s_0, s \leq s_1, \cdots, s \leq s_k\}$ *is non-empty, then it has a unique maximal element.* (Which will be denoted by $s_0 \cap s_1 \cap \cdots \cap s_k$.)

Property (1) implies that for any $s_0, s_1, \cdots, s_k \in S$, $X^{s_0} \cap X^{s_1} \cap \cdots \cap X^{s_k}$ is either empty or of the form X^s $(s = s_0 \cap s_1 \cap \cdots \cap s_k \in S)$.

Let N be the nerve of the maximal covering of X(S).

LEMMA. *If* S *satisfies* (1), *there exists a natural isomorphism* $H_i(S; \mathfrak{g}) \cong H_i(N; N_0(\mathfrak{g}))$, $i \geq 0$, *where* $N(\mathfrak{g})$ *is the coefficient system over* N *defined by* $N_0(\mathfrak{g})(s_0, s_1, \cdots, s_k) = \mathfrak{g}_{s_0 \cap s_1 \cap \cdots \cap s_k}$ *for any* $s_0, s_1, \cdots, s_k \in S^{max}$ *such that* $s_0 \cap s_1 \cap \cdots \cap s_k$ *is defined (the connecting homomorphism of* $N_0(\mathfrak{g})$ *are the obvious ones).*

This follows from the acyclic covering lemma (1.2, $a = \infty$) and 1.4 applied to S^s.

1.6 DEFINITION. *We say that* S *is non-discrete if there exist* $s_1, s_2 \in S$ *such that* $s_1 < s_2$ (*i.e., weight* $(S) \geq 1$).

This is equivalent to the requirement that X(S) is not zero-dimensional. Given $s \in S$ let $\overline{S}^s = \{s' \in S | s' < s\}$. It is clear that $\overline{S}^s = S^s - \{s\}$. We have the following

LEMMA. *Assume* S *is non-discrete and satisfies* (1). *Assume also that* $H_i(S; \mathfrak{g}) = 0$ *for all* i *such that* $0 < i <$ weight (S^s). *Moreover we assume that for all* $s \in S$ *such that* $\overline{S}^s$ *is non-discrete, we have* $H_i(\overline{S}^s; \mathfrak{g}) = 0$ *for all* i *such that* $0 < i <$ weight $(\overline{S}^s)$ *and* $H_0(\overline{S}^s; \mathfrak{g}) \to \mathfrak{g}_s$ *is an isomorphism. Then for any* j, $1 \leq j \leq$ weight (S), *we have* $H_i(S_j; \mathfrak{g}) = 0$ *for all* i *such that* $0 < i < j$ *and* $H_0(S_j; \mathfrak{g}) \to H_0(S; \mathfrak{g})$ *is an isomorphism.*

Proof. First note that the subsets $\overline{S}^s (s \in S)$ and S_j $(1 \leq j \leq \text{weight } (S))$ satisfy (1) if S satisfies (1). Next observe that if $s' \in \overline{S}^s$ we have $(\overline{S}^s)^{s'} = \overline{S}^{s'}$. The lemma is obvious if weight $(S) = 1$. Assume weight $(S) \geq 2$. Since weight $(\overline{S}^s) <$ weight (S), $s \in S$, we can assume by induction on the weight that the result of the lemma is true if S is replaced by $\overline{S}^s$ (where $s \in S$ is such that $\overline{S}^s$ is non-discrete). Hence we can assume that for such s, $H_i((\overline{S}^s)_j; \mathfrak{g}) = 0$, for all i, j such that $0 < i < j \leq$ weight $(\overline{S}^s)$, and $H_0((\overline{S}^s)_j; \mathfrak{g}) \to H_0(\overline{S}^s; \mathfrak{g})$ is an isomorphism (hence also $H_0((\overline{S}^s)_j; \mathfrak{g}) \to \mathfrak{g}_s$ is an isomorphism).

It is easy to see that $(\overline{S}^s)_j = S^s \cap S_j$, if $j \leq$ weight $(\overline{S}^s)$. It follows that $H_i(S^s \cap S_j; \mathfrak{g}) = 0$ for all i, j such that $0 < i < j \leq$ weight $(\overline{S}^s)$ and $H_0(S^s \cap S_j; \mathfrak{g}) \to \mathfrak{g}_s$ is an isomorphism. $(1 \leq j \leq \text{weight } \overline{S}^s.)$

On the other hand, if weight $(S^s) < j \leq$ weight (S) we have clearly $S^s \cap S_j = S^s$. Using Lemma 1.4 we conclude that $H_i(S^s \cap S_j; \mathfrak{g}) = 0$ for all i, j such that $0 < i < j \leq$ weight (S) and $H_0(S^s \cap S_j; \mathfrak{g}) \to \mathfrak{g}_s$ is an isomorphism for $1 \leq j \leq$ weight (S). There we have assumed that S^s is non-discrete, but the result remains valid without this assumption. (The only extra case to be examined is when weight $(s) \leq 1$ in which case $S^s \cap S_j = S^s$.)

These results can be interpreted as follows. The maximal covering of $X(S)$ gives by intersection with $X(S_j)$ a covering of $X(S_j)$ which has clearly the same nerve N as the original covering. Let $N_{j,i}(\mathfrak{g})$ be the coefficient system over N such that $N_{j,i}(\mathfrak{g})(s_0, s_1, \cdots, s_k) = H_i(S^{s_0 \cap s_1 \cap \cdots \cap s_k} \cap S_j; \mathfrak{g})$. We have proved that $N_{j,i}(\mathfrak{g}) = 0$ if $0 < i < j \leq$ weight (S) and $N_{j,0}(\mathfrak{g}) = N_{M,0}(\mathfrak{g})$. $(M = \text{weight } (S),\ 1 \leq j \leq M.)$

We apply now the acyclic covering lemma 1.2 with $a = j$, and we conclude that $H_i(S_j; \mathfrak{g}) \cong H_i(N, N_{M,0}(\mathfrak{g}))$. $0 \leq i < j \leq$ weight $(S) = M$. Take now $j = M$; we get $H_i(N; N_{M,0}(\mathfrak{g})) \cong H_i(S; \mathfrak{g}) = 0$, $0 < i < M$ and $H_0(N; N_{M,0}(\mathfrak{g})) \cong H_0(S; \mathfrak{g})$. It follows that $H_i(S_j; \mathfrak{g}) = 0$, $0 < i < j \leq M$, $H_0(S_j; \mathfrak{g}) = H_0(S; \mathfrak{g})$, $1 \leq j \leq M$ and the lemma is proved by induction.

1.7 We shall now interpret the modules $H_i(S_i, S_{i-1}; \mathfrak{g})$ in terms of chains. The i-chains of S_i are clearly the elements of $C_i(S_i; \mathfrak{g}) = \oplus \mathfrak{g}_{s_0}$ where the sum is over all sequences $s_0 < s_1 < \cdots < s_i$ in S such that weight $(s_0) = 0$, weight $(s_1) = 1, \cdots,$ weight $(s_i) = i$. In other words $C_i(S_i; \mathfrak{g})$ consists of functions which to any sequence $s_0 < s_1 < \cdots < s_i$ as above associate an element in $\mathfrak{g}_{s_0}$. We say that a chain $u \in C_i(S_i; \mathfrak{g})$ satisfies the a-th *cycle condition* $(0 < a \leq i)$ if

$$\sum_{s_a} u(s_0 < s_1 < \cdots s_{a-1} < s_a < s_{a+1} < \cdots < s_i) = 0$$

in $\mathfrak{g}_{s_0}$ for any fixed $s_0 < s_1 < \cdots < s_{a-1} < s_{a+1} < \cdots < s_i$ of weights: $0, 1, \cdots, a-1, a+1, \cdots, i$ and the sum is over all s_a such that $s_{a-1} < s_a < s_{a+1}$ (hence weight $(s_a) = a$). Similarly we say that a chain $u \in C_i(S_i; \mathfrak{g})$ $(i \geq 1)$ satisfies the 0-th *cycle condition* if

$$\sum_{s_0} f_{s_1}^{s_0}\, u(s_0 < s_1 < \cdots < s_i) = 0$$

in $\mathfrak{g}_{s_1}$ for any fixed $s_1 < s_2 < \cdots < s_i$ of weights $1, 2, \cdots, i$ and the sum is over all s_0 such that $s_0 < s_1$ (hence weight $(s_0) = 0$). It is clear that $H_i(S_i; \mathfrak{g}) = \{u \in C_i(S_i; \mathfrak{g}) | u$ satisfies the a-th *cycle condition* for all $a, a = 0, 1, \cdots i\}$. $(i \geq 1)$ and $H_i(S_i, S_{i-1}; \mathfrak{g}) = \{u \in C_i(S_i; \mathfrak{g}) | u$ satisfies the a-th *cycle condition* for all $a, a = 0, 1, \cdots, i-1\}$ $(i \geq 1)$. Moreover $H_0(S_0; \mathfrak{g}) = \bigoplus_{s \in S_0} \mathfrak{g}_s$.

Using these identifications the homomorphisms in the exact sequence of Lemma 1.3 can be described as follows (assume that the hypotheses of Lemma 1.3 are verified). The homomorphism $d: H_i(S_i, S_{i-1}; \mathfrak{g}) \to H_{i-1}(S_{i-1}, S_{i-2}; \mathfrak{g}), (1 \leq i \leq M)$, is described by

$$(du)(s_0 < s_1 < \cdots < s_{i-1}) = \sum_{s_i} u(s_0 < s_1 < \cdots < s_{i-1} < s_i)$$

where the sum is over all s_i of weight i such that $s_{i-1} < s_i$. The remaining homomorphisms (at the two extremes of the exact sequence) are obvious.

Another consequence of this chain description of $H_i(S_i, S_{i-1}; \mathfrak{g})$ is that

$$H_i(S_i, S_{i-1}; \mathfrak{g}) = \bigoplus_{\substack{s \in S \\ \text{weight}(s)=i}} H_{i-1}(\bar{S}^s; \mathfrak{g}), \quad \text{if } i \geq 2$$

and

$$H_i(S_1, S_0; \mathfrak{g}) = \bigoplus_{\substack{s \in S \\ \text{weight}(s)=1}} \ker(H_0(\bar{S}^s; \mathfrak{g}) \to \mathfrak{g}_s)$$

so that the exact sequence of Lemma 3 can be written as

$$0 \to H_M(S; \mathfrak{g}) \to \bigoplus_{\substack{s \in S \\ \text{weight}(s)=M}} H_{M-1}(\bar{S}^s; \mathfrak{g}) \to \bigoplus_{\substack{s \in S \\ \text{weight}(s)=M-1}} H_{M-2}(\bar{S}^s; \mathfrak{g}) \to \cdots$$

$$\to \bigoplus_{\substack{s \in S \\ \text{weight}(s)=2}} H_1(\bar{S}^s; \mathfrak{g}) \to \bigoplus_{\substack{s \in S \\ \text{weight}(s)=1}} \ker(H_0(\bar{S}^s; \mathfrak{g}) \to \mathfrak{g}_s) \to$$

$$\to \bigoplus_{\substack{s \in S \\ \text{weight}(s)=0}} \mathfrak{g}_s \to H_0(S; \mathfrak{g}) \to 0 .$$

1.8 *Examples*

Let E be an affine space of dimension ≥ 2 over a finite field F and let V be any F-vector space of dimension ≥ 2. We shall associate to E and V some partially ordered sets as follows.

Let $S_I = S_I(E)$ be the set of all affine subspaces of E other than E itself, partially ordered by inclusion (if $E', E'' \subset E$ we say that $E' \leq E''$ $\Longleftrightarrow E' \subset E''$). In case E is the underlying affine space of V, i.e., if V is obtained from E by choosing an origin, we shall write $S_I(E) = S_I(V)$.

Let $S_{II} = S_{II}(V)$ be the subset of $S_I(V)$ consisting of those affine subspaces of V which contain 0 but are $\neq 0$ (i.e., the set of all proper linear subspaces of V).

Let $S_{III} = S_{III}(V)$ be the subset of $S_I(V)$ consisting of those affine subspaces of V which do not contain 0.

Let $\tilde{V}$ be a fixed linear subspace of V, $(V \neq 0, V)$. Let $S_{IV} = S_{IV}(V,\tilde{V})$ be the subset of $S_{II}(V)$ consisting of those linear subspaces V' of V which are transversal to $\tilde{V}$ (i.e., $V' + \tilde{V} = V$).

The sets S_{II}, S_{III}, S_{IV} have the partial order induced from that of S_I. Note that S_I and S_{III} are non-discrete if $\dim V \geq 2$, S_{II} is non-discrete if $\dim V \geq 3$; S_{IV} is non-discrete if $\dim V > \dim \tilde{V} \geq 2$. Each of these partially ordered sets satisfies condition (1) of 1.5. For each of these partially ordered sets we can consider a corresponding direct system which is constant and equal to A for each index and whose connecting homomorphisms are the identity maps. This direct system will be denoted again by A. In addition, there is a natural non-constant direct system $\mathcal{G}_{II}$ of F-vector spaces with set of indices S_{II}; $\mathcal{G}_{II}$ is defined by $\mathcal{G}_{II,V'} = V'$ where V' is a proper linear subspace of V; the connecting homomorphisms are given by natural inclusions. Note that $X(S_{II})$ is precisely the Tits complex of the vector space V.

THEOREM (J. Tits-L. Solomon, see [15], Theorem 5.4). $H_i(S_{II}; A) = 0$ *for all* i *such that* $0 < i < n-2$, *and* $H_0(S_{II}; A) \cong A$ *where* $n = \dim V \geq 3$.

This can be proved by a method of Folkman ([2], Theorem 4.1) as follows. Consider the nerve N_{II} of the maximal covering (see 1.5) of $X(S_{II})$. It is enough to prove that $H_i(N_{II}; A) = 0$, $0 < i < n-2$ and $H_0(N_{II}; A) \cong A$. Note that the k-simplices of N_{II} are the sets consisting of $k+1$ linear hyperplanes in V, whose intersection is non-zero. If $k \leq n-2$ this is no restriction, hence the $(n-2)$-skeleton of N_{II} is the same as the $(n-2)$-skeleton of a certain standard simplex and the result follows.

We wish to prove analogous results for the other sets (or direct systems) associated to V.

1.9 THEOREM. $H_i(S_I; A) = 0$ *for all* i *such that* $0 < i < \ell - 1$ *and* $H_0(S_I; A) \cong A$ *where* $\ell = \dim E \geq 2$.

Proof. Note that the homology is not changed if we change temporarily the partial order in S_I to the opposite one (the coefficient system is constant). Using the acyclic covering lemma for the maximal covering with respect to the opposite order we find that it is enough to prove that $H_i(N_I; A) = 0$, $u < i < \ell - 1$. $H_0(N_I; A) \cong A$, where N_I is the simplicial complex whose k-simplices are those subsets of E which have $(k+1)$ elements and which span an affine subspace different from E. If $k \leq \ell - 1$ this is always the case, hence N_I has the same $(\ell-1)$-skeleton as a certain standard simplex and the theorem is proved.

1.10 THEOREM. $H_i(S_{III}; A) = 0$ *for all* i *such that* $0 < i < n-1$ *and* $H_0(S_{III}; A) \cong A$, *where* $n = \dim V \geq 2$.

Just as in the previous proof it is enough to prove the following.

PROPOSITION. *Let* N_{III} *be the simplicial complex whose* k*-simplices are those subsets of* V *which have* $(k+1)$ *elements and which span an affine subspace not containing* 0. *Then* $H_i(N_{III}; A) = 0$ *for all* i *such that* $0 < i < n-1$ *and* $H_0(N_{III}; A) \cong A$, $(n \geq 2)$.

I am indebted to M. Kervaire for supplying for me the following.

Proof of the proposition. First observe that N_{III} is connected. This follows from the fact that given $v_1 \neq v_2$ in V such that $v_1 \neq 0, v_2 \neq 0$, we can find $v_3 \in V$ such that $0, v_1, v_3$ are not collinear and $0, v_2, v_3$ are not collinear. Let $\mathcal{F}$ be the set of all subsets of V, which span an affine subspace not containing 0. Let $Q_1, Q_2, \cdots, Q_n$ be a fixed basis for V. For any ℓ, $1 \leq \ell \leq n$ define a subset $\mathcal{P}_\ell \subset \mathcal{F}$ as follows: $\mathcal{P}_\ell$ consists of all subsets $(v_0, v_1, \cdots, v_k)$ of V such that $\langle v_0, v_1, \cdots, v_k \rangle \not\ni 0$ and $[v_0, v_1, \cdots, v_k] \cap ([Q_1, Q_2, \cdots, Q_{\ell-1}] + Q_\ell) = \emptyset$. Here $\langle\ \rangle$ denotes the affine span, and $[\ \ ]$ denotes the linear span.

We shall consider alternate simplicial chains on N_{III}. Let $C_k(N_{III}; A)$ be the set of k-chains. Given a k-chain u, define its support as

$$\text{supp } u = \{(v_0, v_1, \cdots, v_k) \in \mathcal{F} \mid u(v_0, v_1, \cdots, v_k) \neq 0\} .$$

Define a map

$$T_\ell : \{u \in C_k(N_{III}; A) \mid \text{supp } u \subset \mathcal{P}_\ell\} \to C_{k+1}(N_{III}; A)$$

$(1 \leq \ell \leq n)$ by the formula:

$$(T_\ell u)(Q_\ell, v_0, v_1, \cdots, v_k) = u(v_0, v_1, \cdots, v_k)$$

if $(v_0, v_1, \cdots, v_k) \in \mathcal{P}_\ell$. On sequences which are permutations of $(Q_\ell, v_0, v_1, \cdots, v_k)$ define $T_\ell u$ using the alternacy conditions; on all other sequences define $T_\ell u$ to be zero.

Note that $(v_0, v_1, \cdots, v_k) \in \mathcal{P}_\ell$ implies $(Q_\ell, v_0, v_1, \cdots, v_k) \in \mathcal{F}$. It is easy to check that if $k \geq 1$, $\partial(T_\ell u) + T_\ell(\partial u) = u$ where u is in the source of T_ℓ (note that $\text{supp } u \subset \mathcal{P}_\ell \Longrightarrow \text{supp } \partial u \subset \mathcal{P}_\ell$).

LEMMA. *Let* $u \in C_k(N_{III}; A)$ *be such that* $\partial u = 0$ *and* $\text{supp } u \subset \mathcal{P}_\ell \cap \mathcal{P}_{\ell+1} \cap \cdots \cap \mathcal{P}_n$. *Then there exist* $u_1 \in C_k(N_{III}; A)$ *and* $u_2 \in C_{k+1}(N_{III}; A)$ *such that* $\text{supp } u_1 \subset \mathcal{P}_{\ell+1} \cup \mathcal{P}_{\ell+2} \cup \cdots \cup \mathcal{P}_n$ *and* $u - u_1 = \partial u_2$ $(k \geq 1)$.

Proof. We can write $u = u' - u''$, where $u', u'' \in C_k(N_{III}; A)$ and $\text{supp } u' \subset \mathcal{P}_\ell$, $\text{supp } u'' \subset \mathcal{P}_{\ell+1} \cup \cdots \cup \mathcal{P}_n$. We have $z = \partial u' = \partial u''$ and $\text{supp } z \subset \mathcal{P}_\ell \cap (\mathcal{P}_{\ell+1} \cup \cdots \cup \mathcal{P}_n)$. Put $u_1 = T_\ell z - u''$, $u_2 = T_\ell u'$. Then

$$\partial u_2 = \partial(T_\ell u') = u' - T_\ell(\partial u') = u' - T_\ell z = u' - u'' + u'' - T_\ell z = u - u_1 .$$

So it is enough to prove that

$$\text{supp } T_\ell z \subset \mathcal{P}_{\ell+1} \cup \cdots \cup \mathcal{P}_n .$$

Assume that $(T_\ell z)(v_0, v_1, \cdots, v_k) \neq 0$. Then we must have (up to a permutation) $v_0 = Q_\ell$, and $(v_1, v_2, \cdots, v_k) \in \operatorname{supp} z \subset \mathcal{P}_{\ell+1} \cup \cdots \cup \mathcal{P}_n$. It follows that $[v_1, v_2, \cdots, v_k] \cap ([Q_1, \cdots, Q_{\ell+h-1}] + Q_{\ell+h}) = \emptyset$ for some $h, 1 \leq h \leq n-\ell$. Hence also $[Q_\ell, v_1, v_2, \cdots, v_k] \cap ([Q_1, \cdots, Q_{\ell+h-1}] + Q_{\ell+h}) = \emptyset$ for some h, $1 \leq h \leq n-\ell$, which implies that $(Q_\ell, v_1, v_2, \cdots, v_k) \in \mathcal{P}_{\ell+1} \cup \cdots \cup \mathcal{P}_n$ and the lemma is proved.

Applying the Lemma repeatedly we find that given $u \in C_k(N_{III}; A)$ such that $\partial u = 0$ and $\operatorname{supp} u \subset \mathcal{P}_1 \cup \mathcal{P}_2 \cup \cdots \cup \mathcal{P}_n$ there exists $u_2 \in C_{k+1}(N_{III}; A)$ such that $u = \partial u_2$ $(k \geq 1)$. Hence to conclude the proof of the proposition it is enough to prove that any k-simplex of N_{III} with $k \leq n-2$ lies in $\mathcal{P}_1 \cup \mathcal{P}_2 \cup \cdots \cup \mathcal{P}_n$. Otherwise, we could find a subset $(v_0, v_1, \cdots, v_k)$ of V such that $[v_0, v_1, \cdots, v_k] \cap ([Q_1, \cdots, Q_{\ell-1}] + Q_\ell) \neq \emptyset$ for all ℓ, $1 \leq \ell \leq n$. It follows by induction that $Q_1, Q_2, \cdots, Q_n \in [v_0, v_1, \cdots, v_k]$ hence $V = [v_0, v_1, \cdots, v_k]$. This contradicts the assumption that $k \leq n-2$ and the proposition is proved.

1.11 THEOREM. *Given a linear subspace* $\tilde{V} \subset V$ *we have* $H_i(S_{IV}; A) = 0$ *for all* i *such that* $0 < i < m-1$ *and* $H_0(S_{IV}; A) \cong A$, *where* $n = \dim V > \dim \tilde{V} = m \geq 2$.

Proof. Using the acyclic covering lemma applied to the maximal covering we see that it is enough to prove that $H_i(N_{IV}; A) = 0$, $0 < i < m-1$ and $H_0(N_{IV}; A) \cong A$, where N_{IV} is the simplicial complex defined as follows: The k-simplices of N_{IV} are precisely the sets $(H_0, H_1, \cdots, H_k)$ of $(k+1)$ linear hyperplanes in V such that $(H_0 \cap H_1 \cap \cdots \cap H_k) + \tilde{V} = V$.

N_{IV} can be also described as the simplicial complex whose k-simplices are the sets $(L_0, L_1, \cdots, L_k)$ of $(k+1)$ one-dimensional linear subspaces of V^* such that $(L_0 + L_1 + \cdots + L_k) \cap \tilde{V}^\perp = 0$ where $\tilde{V}^\perp \subset V^*$ is the orthogonal complement of $\tilde{V}$. Note that N_{IV} is connected (the proof similar to the one in 1.10 uses the fact that $\operatorname{codim} \tilde{V}^\perp \geq 2$).

Let $\mathcal{F}'$ be the set of all sets $(L_0, L_1, \cdots, L_k)$ of one dimensional linear subspaces of V^* such that $(L_0 + L_1 + \cdots + L_k) \cap \tilde{V}^{\perp} = 0$. Let $Q'_1, Q'_2, \cdots, Q'_m$ be a set of vectors in V^* whose images in $V^*/\tilde{V}$ form a basis. For any $\ell, 1 \leq \ell \leq m$, define a subset $\mathcal{P}'_\ell \subset \mathcal{F}'$ as follows: $\mathcal{P}'_\ell$ consists of all elements $(L_0, L_1, \cdots, L_k)$ of $\mathcal{F}'$ such that $(L_0 + L_1 + \cdots + L_k) \cap ([Q'_1, Q'_2, \cdots, Q'_{\ell-1}] + Q'_\ell + \tilde{V}^{\perp}) = \emptyset$. Define a map

$$T'_\ell : \{u \in C_k(N'; A) \mid \operatorname{supp} u \subset \mathcal{P}'_\ell\} \to C_{k+1}(N'; A)$$

by the formula

$$(T'_\ell u)([Q'_\ell], L_0, L_1, \cdots, L_k) = u(L_0, L_1, \cdots, L_k)$$

if $(L_0, L_1, \cdots, L_k) \in \mathcal{P}'_\ell$. On sequences which are permutations of $([Q'_\ell], L_0, L_1, \cdots, L_k)$ define $T'_\ell u$ using the alternacy condition; on other sequences define $T'_\ell u$ to be zero. Note that $(L_0, L_1, \cdots, L_k) \in \mathcal{P}'_\ell \Longrightarrow ([Q'_\ell], L_0, L_1, \cdots, L_k) \in \mathcal{F}'$. One verifies that for $k \geq 1$ we have $\partial(T'_\ell u) + T'_\ell(\partial u) = u$, where u is in the source of T'_ℓ.

LEMMA. *Let* $u \in C_k(N_{IV}; A)$ *be such that* $\partial u = 0$ *and* $\operatorname{supp} u \subset \mathcal{P}'_\ell \cup \mathcal{P}'_{\ell+1} \cup \cdots \cup \mathcal{P}'_m$. *Then there exist* $u_1 \in C_k(N_{IV}; A)$ *and* $u_2 \in C_{k+1}(N_{IV}; A)$ *such that* $\operatorname{supp} u_1 \subset \mathcal{P}'_{\ell+1} \cup \mathcal{P}'_{\ell+2} \cup \cdots \cup \mathcal{P}'_m$ *and* $u - u_1 = \partial u_2$ $(k \geq 1)$.

Proof. Write $u = u' - u''$, where $u', u'' \in C_k(N_{IV}; A)$ and $\operatorname{supp} u' \subset \mathcal{P}'_\ell$, $\operatorname{supp} u'' \subset \mathcal{P}'_{\ell+1} \cup \cdots \cup \mathcal{P}'_m$. Then $z = \partial u' = \partial u''$ satisfies $\operatorname{supp} z \subset \mathcal{P}'_\ell \cap (\mathcal{P}'_{\ell+1} \cup \cdots \cup \mathcal{P}'_m)$. Put $u_1 = T'_\ell z - u''$, $u_2 = T'_\ell u'$. Then $\partial u_2 = u - u_1$, and it is enough to prove that $\operatorname{supp} T'_\ell z \subset \mathcal{P}'_{\ell+1} \cup \cdots \cup \mathcal{P}'_m$. Assume that $(T'_\ell z)(L_0, L_1, \cdots, L_k) \neq 0$. Then we must have (up to a permutation) $L_0 = [Q'_\ell]$, and $(L_1, \cdots, L_k) \in \operatorname{supp} z \subset \mathcal{P}'_{\ell+1} \cup \cdots \cup \mathcal{P}'_m$. It follows that $(L_1 + \cdots + L_k) \cap ([Q'_1, \cdots, Q'_{\ell+h-1}] + Q'_{\ell+h} + \tilde{V}^{\perp}) = \emptyset$ for some h, $1 \leq h \leq m - \ell$. Hence also

$$([Q'_\ell] + L_1 + \cdots + L_k) \cap ([Q'_1, \cdots, Q'_{\ell+h-1}] + Q'_{\ell+h} + \tilde{V}^\perp) = \emptyset$$

for some h, $1 \leq h \leq m-\ell$, which implies that

$$([Q'_\ell], L_1, L_2, \cdots, L_k) \in \mathcal{P}'_{\ell+1} \cup \cdots \cup \mathcal{P}'_m$$

and the lemma is proved.

Applying the lemma repeatedly we find that given $u \in C_k(N_{IV}; A)$ such that $\partial u = 0$ and $\operatorname{supp} u \subset \mathcal{P}'_1 \cup \mathcal{P}'_2 \cup \cdots \cup \mathcal{P}'_m$, there exists $u_2 \in C_{k+1}(N_{IV}; A)$ such that $u = \partial u_2$ $(k \geq 1)$. To conclude the proof of the theorem it is enough to observe that every k-simplex of N_{IV} with $k \leq m-2$ lies in $\mathcal{G}'_1 \cup \mathcal{P}'_2 \cup \cdots \cup \mathcal{P}'_m$. Otherwise, we could find $k+1$ lines in V^* such that

$$(L_0 + L_1 + \cdots + L_k) \cap ([Q'_1, Q'_2, \cdots, Q'_{\ell-1}] + Q'_\ell + \tilde{V}^\perp) \neq \emptyset$$

for all ℓ, $\ell = 1, 2, \cdots, m$. It follows by induction that $Q'_1, Q'_2, \cdots, Q'_m \in L_0 + L_1 + \cdots + L_k + \tilde{V}^\perp$. Since clearly $\tilde{V}^\perp \subset L_0 + L_1 + \cdots + L_k + \tilde{V}^\perp$ it follows that $V^* = L_0 + L_1 + \cdots + L_k + \tilde{V}$ which contradicts the hypothesis $k \leq m-2$ $(m = \operatorname{codim} \tilde{V}^\perp)$ and the theorem is proved.

Remark. This theorem contains Theorem 1.9 as a special case. Replace V by $V \oplus F$ and take $\tilde{V} = V \subset V \oplus F$ $(\operatorname{codim} \tilde{V} = 1)$. Then $S_I(V) = S_{IV}(V \oplus F, V)$. The reason we have proved Theorem 1.9 separately is that its proof is easier than that of the present theorem.

1.12 THEOREM. $H_i(S_{II}; \mathcal{G}_{II}) = 0$ *for all* i *such that* $0 < i < n-2$ *and* $H_0(S_{II}; \mathcal{G}_{II}) \cong V$, *where* $n = \dim V \geq 3$.

Proof. Using the acylic covering lemma applied to the maximal covering of S_{II} (see 1.5) we see that it is enough to prove that $H_i(N_{II}; \tilde{\mathcal{G}}_{II}) = 0$, $0 < i < n-2$ and $H_0(N_{II}; \tilde{\mathcal{G}}_{II}) \cong V$ where N_{II} is the simplicial complex defined in 1.8 and $\tilde{\mathcal{G}}_{II}$ is a coefficient system over N_{II} defined as follows. Given a k-simplex $(H_0, H_1, \cdots, H_k)$ in N_{II} (H_i linear hyperplanes in V such that $H_0 \cap H_1 \cap \cdots \cap H_k \neq 0$) we have $\tilde{\mathcal{g}}_{II}(H_0, H_1, \cdots, H_k) = H_0 \cap H_1 \cap \cdots \cap H_k$. The connecting homomorphisms are given by the natural inclusions. Let $\overline{N}$ be the simplicial complex whose k-simplices are the sets $(H_0, H_1, \cdots, H_k)$ of k linear hyperplanes in V whose intersection is arbitrary (possibly zero). We have $N_{II} \subset \overline{N}$ and $\overline{N}$ is clearly a standard simplex. The coefficient system $\tilde{\mathcal{g}}_{II}$ can be extended to a coefficient system $\overline{\mathcal{g}}$ over $\overline{N}$ such that $\overline{\mathcal{g}} = \tilde{\mathcal{g}}_{II}$ on the simplices of N_{II} and $\overline{\mathcal{g}} = 0$ on the other simplices. We have then clearly $H_i(N_{II}; \tilde{\mathcal{g}}_{II}) \cong H_i(\overline{N}; \overline{\mathcal{g}})$ for all $i \geq 0$. Let $\overline{V}$ denote the constant coefficient system over $\overline{N}$ with value V at every simplex of $\overline{N}$. Then clearly $H_i(\overline{N}; \overline{V}) = 0$ $(i > 0)$ and $H_0(\overline{N}; \overline{V}) \cong V$ since $\overline{N}$ is a standard simplex. There is a natural embedding of coefficient systems $\overline{\mathcal{g}} \subset \overline{V}$ and the quotient $\mathcal{S} = \overline{V}/\overline{\mathcal{g}}$ is defined. We have an exact homology sequence:

$$H_{n-2}(\overline{N}; \mathcal{S}) \to H_{n-3}(\overline{N}; \overline{\mathcal{g}}) \to H_{n-3}(N; \overline{V}) \to \cdots \to H_1(\overline{N}; \mathcal{S}) \to H_0(\overline{N}; \overline{\mathcal{g}})$$
$$\to H_0(\overline{N}; \overline{V}) \to H_0(\overline{N}; \mathcal{S})$$

which shows that it is enough to prove that $H_i(\overline{N}; \mathcal{S}) = 0$ for $0 \leq i \leq n-2$. It will be useful to regard linear hyperplanes in V as one-dimensional linear subspaces in V^*. With this identification, $\overline{N}$ and $\mathcal{S}$ can be described as follows: $\overline{N}$ is the simplicial complex whose k-simplices are the sets $(L_0, L_1, \cdots, L_k)$ of one-dimensional linear subspaces of V^*; we have $\mathcal{S}_{(L_0, L_1, \cdots, L_k)} = (L_0 + L_1 + \cdots + L_k)^*$ and the connecting homomorphisms of $\mathcal{S}$ are obtained by restricting linear functions.

PROPOSITION. $H_i(\overline{N}; \mathcal{S}) = 0$, $0 \leq i \leq n-2$ $(n \geq 3)$.

Proof. This will again be similar to the proof of Proposition 1.10. Let $\overline{\mathcal{F}}$ be the set of all simplices of $\overline{N}$. For any ℓ, $1 \leq \ell \leq n$, define a subset $\overline{\mathcal{P}}_\ell \subset \overline{\mathcal{F}}$ as the set of simplices $(L_0, L_1, \cdots, L_k) \in \overline{\mathcal{F}}$ such that

$L_0+L_1+\cdots+L_k) \cap ([\bar{Q}_1, \bar{Q}_2, \cdots, \bar{Q}_{\ell-1}]+\bar{Q}_\ell) = \emptyset$. Here $\bar{Q}_1, \bar{Q}_2, \cdots, \bar{Q}_n$ is a fixed basis of V^*. There is a natural notion of support for chains in $C_k(\bar{N}; \mathcal{S})$ (we use alternating simplicial chains). Define a map

$$\bar{T}_\ell : \{u \in C_k(\bar{N}; \mathcal{S}) \mid \operatorname{supp} u \subset \bar{\mathcal{P}}_\ell\} \to C_{k+1}(\bar{N}; \mathcal{S})$$

by the formula

$$(\bar{T}_\ell u)([\bar{Q}_\ell], L_0, L_1, \cdots, L_k) = \Pi(u(L_0, L_1, \cdots, L_k))$$

if $(L_0, L_1, \cdots, L_k) \in \bar{\mathcal{P}}_\ell$. On sequences which are permutations of $([\bar{Q}_\ell], L_0, L_1, \cdots, L_k)$ define $\bar{T}_\ell u$ using the alternacy condition; on other sequences define $\bar{T}_\ell u$ to be zero. (Here Π is the natural linear map $\Pi : (L_0+\cdots+L_k)^* \to ([\bar{Q}_\ell]+L_0+\cdots+L_k)^*$ defined as the dual of the natural projection $[\bar{Q}_\ell] \oplus (L_0+L_1+\cdots+L_k) \to L_0+L_1+\cdots+L_k$; note that the sum $[\bar{Q}_\ell]+(L_0+L_1+\cdots+L_k)$ is direct since $(L_0, L_1, \cdots, L_k) \in \bar{\mathcal{P}}_\ell$.)

We shall verify that for all $k \geq 0$ we have

$$\partial(\bar{T}_\ell u) + \bar{T}_\ell(\partial u) = u, \quad \text{where } u \in C_k(N; \mathcal{S}) \text{ and } \operatorname{supp} u \subset \bar{\mathcal{P}}_\ell. \tag{2}$$

In fact, let $(L_0, L_1, \cdots, L_k) \in \bar{\mathcal{P}}_\ell$; we have

$$\begin{aligned}\partial(\bar{T}_\ell u)(L_0, L_1, \cdots, L_k) &= \phi((\bar{T}_\ell u)([\bar{Q}_\ell], L_0, L_1, \cdots, L_k)) \\ &= \phi\Pi\, u(L_0, L_1, \cdots, L_k) = u(L_0, L_1, \cdots, L_k)\end{aligned}$$

where $\phi : ([\bar{Q}_\ell]+L_0+\cdots+L_k)^* \to (L_0+\cdots+L_k)^*$ is given by restricting linear functions. We also have in this case $\bar{T}_\ell(\partial u)(L_0, L_1, \cdots, L_k) = 0$, hence the equality (2) is checked on simplices $(L_0, L_1, \cdots, L_k) \in \bar{\mathcal{P}}_\ell$. For other simplices, the right hand side of (2) is zero. If $(L_0, L_1, \cdots, L_k) \notin \bar{\mathcal{P}}_\ell$ and $[\bar{Q}_\ell] \neq L_0, L_1, \cdots, L_k$, all terms of (2) are zero. Hence it is enough to check that the left hand side of (2) is zero on simplices of the form $([\bar{Q}_\ell], L_1, L_2, \cdots, L_k)$.

We have

$$\partial(\overline{T}_\ell u)([\overline{Q}_\ell], L_1, L_2, \cdots, L_k) = -\sum_{L} \phi_1 \Pi_1 \; u(L, L_1, \cdots, L_k)$$

where the sum is over all L such that $L \neq L_1, L_2, \cdots, L_k$, $(L, L_1, \cdots, L_k) \in \overline{\mathcal{P}}_\ell$, and

$$\overline{T}_\ell(\partial u)([\overline{Q}_\ell], L_1, L_2, \cdots, L_k) = \sum_{L} \Pi_2 \, \phi_2 \; u(L, L_1, \cdots, L_k)$$

where L runs over the same set as above. Here $\phi_1, \phi_2, \Pi_1, \Pi_2$ are the arrows in the dual of the commutative diagram:

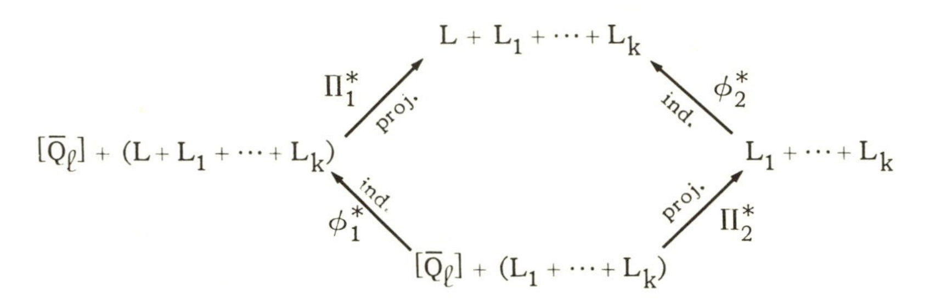

It follows that $\phi_1 \Pi_1 = \Pi_2 \phi_2$ and the equality (2) is proved. Note that in contrast to the analogous equalities in 1.10, 1.11, the equality (2) holds even for $k = 0$.

LEMMA. *Let* $u \in C_k(\overline{N}; \mathcal{S})$ *be such that* $\partial u = 0$ *and* $\operatorname{supp} u \subset \overline{\mathcal{P}}_\ell \cup \overline{\mathcal{P}}_{\ell+1} \cup \cdots \cup \overline{\mathcal{P}}_n$. *Then there exist* $u_1 \in C_k(\overline{N}; \mathcal{S})$ *and* $u_2 \in C_{k+1}(\overline{N}; \mathcal{S})$ *such that* $\operatorname{supp} u_1 \subset \overline{\mathcal{P}}_{\ell+1} \cup \cdots \cup \overline{\mathcal{P}}_n$ *and* $u - u_1 = \partial u_2$ $(k \geq 0)$.

Proof. Write $u = u' - u''$, where $u', u' \in C_k(\overline{N}; \mathcal{S})$ and $\operatorname{supp} u' \subset \overline{\mathcal{P}}_\ell$, $\operatorname{supp} u'' \subset \overline{\mathcal{P}}_{\ell+1} \cup \cdots \cup \overline{\mathcal{P}}_n$. Then $z = \partial u' = \partial u''$ satisfies $\operatorname{supp} z \subset \overline{\mathcal{P}}_\ell \cap (\overline{\mathcal{P}}_{\ell+1} \cup \cdots \cup \overline{\mathcal{P}}_n)$. Put $u_1 = \overline{T}_\ell z - u''$, $u_2 = \overline{T}_\ell u'$. Then $\partial u_2 = u - u_1$ and it is enough to prove that $\operatorname{supp} \overline{T}_\ell z \subset \overline{\mathcal{P}}_{\ell+1} \cup \cdots \cup \overline{\mathcal{P}}_n$. Assume that $(\overline{T}_\ell z)(L_0, L_1, \cdots, L_k) \neq 0$. Then we must have up to a permutation $L_0 = [\overline{Q}_\ell]$ and $(L_1, \cdots, L_k) \in \operatorname{supp} z \subset \overline{\mathcal{P}}_{\ell+1} \cup \cdots \cup \overline{\mathcal{P}}_n$. Hence $(L_1 + \cdots + L_k) \cap$

$[\overline{Q}_1, \overline{Q}_2, \cdots, \overline{Q}_{\ell+h-1}] + \overline{Q}_{\ell+h} = \emptyset$ for some h, $1 \leq h \leq n-\ell$. It follows that $([\overline{Q}_\ell] + L_1 + \cdots + L_k) \cap [\overline{Q}_1, \overline{Q}_2, \cdots, \overline{Q}_{\ell+h-1}] + \overline{Q}_{\ell+h} = \emptyset$ for some h, $1 \leq h \leq n-\ell$. Hence $([\overline{Q}_\ell], L_1, L_2, \cdots, L_k) \in \overline{\mathcal{P}}_{\ell+1} \cup \cdots \cup \overline{\mathcal{P}}_n$ and the lemma is proved.

Applying the lemma repeatedly we find that given $u \in C_k(\overline{N}; \mathcal{S})$ such that $\partial u = 0$ and $\operatorname{supp} u \subset \overline{\mathcal{P}}_1 \cup \overline{\mathcal{P}}_2 \cup \cdots \cup \overline{\mathcal{P}}_n$ there exists $u_2 \in C_{k+1}(\overline{N};\mathcal{S})$ such that $u = \partial u_2$ $(k \geq 0)$. To conclude the proof of the proposition it is hence enough to prove that any k-simplex of $\overline{N}$, with $k \leq n-2$ lies in $\overline{\mathcal{P}}_1 \cup \overline{\mathcal{P}}_2 \cup \cdots \cup \overline{\mathcal{P}}_n$. Otherwise, we could find $(k+1)$ lines $L_0, L_1, \cdots, L_k$ in V^* such that $(L_0 + L_1 + \cdots + L_k) \cap [\overline{Q}_1, \overline{Q}_2, \cdots, \overline{Q}_{\ell-1}] + \overline{Q}_\ell \neq \emptyset$ for all ℓ, $1 \leq \ell \leq n$. It follows by induction that $\overline{Q}_1, \overline{Q}_2, \cdots, \overline{Q}_n \in L_0 + L_1 + \cdots + L_k$ and hence $V^* = L_0 + L_1 + \cdots + L_k$ which contradicts the assumption that $k \leq n-2$. The proposition is proved.

1.13 It follows from 1.8 - 1.12 that all the pairs $(S; \mathcal{g}) = (S_I; A)$, $(S_{II}; A)$, $(S_{II}, \mathcal{g}_{II})$, $(S_{III}; A)$, $(S_{IV}; A)$ have non-zero homology only in the top and bottom dimension. Moreover direct inspection shows that all the subsets $\overline{S}^s$ of S ($s \in S$ such that $\overline{S}^s$ non-discrete) together with the restriction of $\mathcal{g}$ to $\overline{S}^s$ are again in this list, possibly for a smaller E, V or $\tilde{V}$. It follows that the hypothesis of Lemma 1.6 are satisfied by each of these pairs, and hence also the hypothesis of Lemma 1.3 are satisfied. We can then write the exact sequence of 1.7 for each of these pairs. But first we introduce some notation.

If E is some affine space of over F dimension $\ell \geq 2$ we define $\Delta_A(E) = H_{\ell-1}(S_I(E);A)$. If $\dim E = 1$ we define $\Delta_A(E) = \ker(H_0(S_I(E); A) \to A)$ where the arrow is the obvious augmentation homomorphism; if $\dim E = 0$ we define $\Delta_A(E) = A$.

Now let V be a vector space over F of dimension $n \geq 3$. Define $St_A(V) = H_{n-2}(S_{II}(V); A)$ (Steinberg module) and $\mathcal{D}(V) = H_{n-2}(S_{II}(V); \mathcal{g}_{II})$. If $\dim V = 2$, define $St_A(V) = \ker(H_0(S_{II}(V); A) \to A)$ and $\mathcal{D}(V) = \ker(H_0(S_{II}(V); \mathcal{g}_{II}) \to V)$. Here the arrows are the obvious augmentation homomorphisms (obtained by summing over all lines in V).

If $\dim V = 1$, define $St_A(V) = A$ and $\mathfrak{D}(V) = V$. Assume now that $\dim V \geq 2$; define $\dot{\Delta}_A(V) = H_{n-1}(S_{III}(V); A)$. If $\dim V = 1$ define $\dot{\Delta}_A(V) = \ker (H_0(S_{III}(V); A) \to A)$.

Let now $\tilde{V} \subset V$ be such that $n = \dim V > \dim \tilde{V} = m \geq 2$. Define $P_A(V, \tilde{V}) = H_{m-1}(S_{IV}(V, \tilde{V}); A)$. If $\dim V > \dim \tilde{V} = 1$ define

$$P_A(V, \tilde{V}) = \ker (H_0(S_{IV}(V, \tilde{V}); A) \to A) ;$$

if $\dim V > \dim \tilde{V} = 0$ define

$$P_A(V, \tilde{V}) = A .$$

We can now write the exact sequence of 1.7 in each case:

(a) $$0 \to \Delta_A(E) \to \bigoplus_{E_{\ell-1}} \Delta_A(E_{\ell-1}) \to \bigoplus_{E_{\ell-2}} \Delta_A(E_{\ell-2}) \to \cdots \to \bigoplus_{E_1} \Delta_A(E_1)$$
$$\to \bigoplus_{E_0} \Delta_A(E_0) \to A \to 0 .$$

Here E is an affine space over F of dimension $\ell \geq 1$ and $\bigoplus_{E_i}$ means direct sum over all i-dimensional affine subspaces of E.

(b) $$0 \to St_A(V) \to \bigoplus_{V_{n-1}} St_A(V_{n-1}) \to \bigoplus_{V_{n-2}} St_A(V_{n-2}) \to \cdots \to \bigoplus_{V_2} St_A(V_2)$$
$$\to \bigoplus_{V_1} St_A(V_1) \to A \to 0$$

(c) $$0 \to \mathfrak{D}(V) \to \bigoplus_{V_{n-1}} \mathfrak{D}(V_{n-1}) \to \bigoplus_{V_{n-2}} \mathfrak{D}(V_{n-2}) \to \cdots \to \bigoplus_{V_2} \mathfrak{D}(V_2)$$
$$\to \bigoplus_{V_1} \mathfrak{D}(V_1) \to V \to 0 .$$

In case (b) and (c), V is a vector space over F of dimension $n \geq 2$ and $\bigoplus_{V_i}$ means direct sum over all i-dimensional linear subspaces of V.

(d) $$0 \to \dot{\Delta}_A(V) \to \bigoplus_{E_{n-1} \not\ni 0} \dot{\Delta}_A(E_{n-1}) \to \bigoplus_{E_{n-2} \not\ni 0} \dot{\Delta}_A(E_{n-2}) \to \cdots$$

$$\to \bigoplus_{E_1 \not\ni 0} \dot{\Delta}_A(E_1) \to \bigoplus_{E_0 \not\ni 0} \dot{\Delta}_A(E_0) \to A \to 0 \; .$$

Here V is a vector space over F of dimension $n \geq 1$ and $\bigoplus_{E_i \not\ni 0}$ means direct sum over all i-dimensional affine subspaces of V which do not contain 0.

(e) $$0 \to P_A(V,\tilde{V}) \to \bigoplus_{V_{n-1} \pitchfork \tilde{V}} P_A(V_{n-1}, V_{n-1} \cap \tilde{V}) \to \bigoplus_{V_{n-2} \pitchfork \tilde{V}} P_A(V_{n-2}, V_{n-2} \cap \tilde{V})$$

$$\to \cdots \to \bigoplus_{V_{n-m+1} \pitchfork \tilde{V}} P_A(V_{n-m+1}, V_{n-m+1} \cap \tilde{V}) \to \bigoplus_{V_{n-m} \pitchfork \tilde{V}} P_A(V_{n-m}, 0)$$

$$\to A \to 0 \; .$$

Here $\tilde{V} \subset V$ and $n = \dim V > \dim \tilde{V} = m \geq 1$, and the sum $\bigoplus_{V_i \pitchfork \tilde{V}}$ is the direct sum over all i-dimensional linear subspaces V_i of V such that $V_i + \tilde{V} = V$. (The symbol $\pitchfork$ means: transversal.) Note that the homomorphisms in these exact sequences can be described explicitly in terms of chains, as explained in 1.7.

1.14 *Dimension formulae*

We recall that A is an arbitrary commutative ring. From the results of 1.8 - 1.12 and the universal coefficient formula it follows that the simplicial complexes S_I, S_{II}, S_{III}, S_{IV} are torsion free and we have: $\Delta_A(E) = \Delta_{\mathbf{Z}}(E) \otimes A$, $St_A(V) = St_{\mathbf{Z}}(V) \otimes A$, $\dot{\Delta}_A(V) = \dot{\Delta}_{\mathbf{Z}}(V) \otimes A$, $P_A(V, \tilde{V}) = P_{\mathbf{Z}}(V, \tilde{V}) \otimes A$. It follows also that $\Delta_A(E)$, $St_A(V)$, $\dot{\Delta}_A(V)$, $P_A(V, \tilde{V})$ are free A-modules of finite rank, independent of A. Note that $\mathfrak{D}(V)$ is an F-vector space of finite dimension. The next theorem describes the ranks of these modules.

THEOREM. *Let* E *be an affine space over* F, V *a vector space over* F *and* $\tilde{V}$ *a linear subspace of* V. *Let* q *be the number of elements in* F. *Then*

$$\operatorname{rank}_A \Delta_A(E) = (q-1)(q^2-1)\cdots(q^\ell-1), \quad (\dim E = \ell \geq 1)$$

$$\operatorname{rank}_A \mathrm{St}_A(V) = q\cdot q^2 \cdots q^{n-1}, \qquad (\dim V = n \geq 2)$$

$$\dim_F \mathcal{D}(V) = (q-1)(q^2-1)\cdots(q^{n-1}-1), (\dim V = n \geq 2)$$

$$\operatorname{rank}_A \dot{\Delta}_A(V) = \sum_{i=1}^{n} (-1)^{i-1}(q^i-1)(q^{i+1}-1)\cdots(q^n-1) + (-1)^n, \quad (\dim V = n \geq 1)$$

$$\operatorname{rank}_A P_A(V,\tilde{V}) = (q^{n-m}-1)(q^{n-m+1}-1)\cdots(q^{n-1}-1) \quad (\dim V = n > \dim \tilde{V} = m \geq 1).$$

Proof. Let

$$\alpha(\ell) = \operatorname{rank}_A \Delta_A(E), \quad \beta(n) = \operatorname{rank}_A \mathrm{St}_A(V),$$

$$\gamma(n) = \dim_F \mathcal{D}(V), \qquad \delta(n) = \operatorname{rank}_A \dot{\Delta}_A(V),$$

$\varepsilon(n,m) = \operatorname{rank}_A P_A(V,\tilde{V})$, where E, V, $\tilde{V}$ are as in the Theorem. We can find the values of $\alpha(\ell)$, $\beta(n)$, $\gamma(n)$, $\delta(n)$, $\varepsilon(n,m)$ by induction from the exact sequences 1.13. We shall first determine $\alpha(\ell)$. It is easy to see that the number of i-dimensional affine subspaces of E equals

$$q^{\ell-i}\frac{(q^{\ell-i+1}-1)(q^{\ell-i+2}-1)\cdots(q^\ell-1)}{(q-1)(q^2-1)\cdots(q^i-1)}.$$

It then follows from 1.13(a) that

$$\alpha(\ell) + \sum_{i=0}^{\ell-1}(-1)^{\ell-i}\alpha(i)q^{\ell-i}\frac{(q^{\ell-i+1}-1)(q^{\ell-i+2}-1)\cdots(q^\ell-1)}{(q-1)(q^2-1)\cdots(q^i-1)} + (-1)^{\ell+1} = 0$$

where $\alpha(0) = 1$ $(\ell \geq 1)$. It is clear that $\alpha(1) = q-1$. Assume that $\alpha(i) = (q-1)(q^2-1)\cdots(q^i-1)$ for $1 \leq i \leq \ell-1$. Substituting this value of $\alpha(i)$ we get

$$\alpha(\ell) + \sum_{i=0}^{\ell-1}(-1)^{\ell-i}q^{\ell-i}(q^{\ell-i+1}-1)(q^{\ell-i+2}-1)\cdots(q^\ell-1) + (-1)^{\ell+1} = 0$$

or

$$(-1)^{\ell}\alpha(\ell) = 1 - \sum_{i=0}^{\ell-1} q^{\ell-i}(1-q^{\ell-i+1})(1-q^{\ell-i+2})\cdots(1-q^{\ell}) .$$

Replace now the factors $q^{\ell-i}$ by $1-(1-q^{\ell-i})$; we get

$$(-1)^{\ell}\alpha(\ell) = 1 + \sum_{i=0}^{\ell-1}(1-q^{\ell-i})(1-q^{\ell-i+1})\cdots(1-q^{\ell})$$

$$- \sum_{i=0}^{\ell-1}(1-q^{\ell-i+1})(1-q^{\ell-i+2})\cdots(1-q^{\ell})$$

$$= 1 - 1 + (1-q)(1-q^2)\cdots(1-q^{\ell}) .$$

Hence $\alpha(\ell) = (q-1)(q^2-1)\cdots(q^{\ell}-1)$.

Next, observe that the number of i-dimensional linear subspaces of V equals:

$$\frac{(q^{n-i+1}-1)(q^{n-i+2}-1)\cdots(q^n-1)}{(q-1)(q^2-1)\cdots(q^i-1)} .$$

It follows from 1.13 (b) and 1.13 (c) that for any $n \geq 2$:

$$\beta(n) + \sum_{i=1}^{n-1}(-1)^{n-i}\beta(i)\ \frac{(q^{n-i+1}-1)(q^{n-i+2}-1)\cdots(q^n-1)}{(q-1)(q^2-1)\cdots(q^i-1)} + (-1)^n = 0$$

$$\gamma(n) + \sum_{i=1}^{n-1}(-1)^{n-i}\gamma(i)\ \frac{(q^{n-i+1}-1)(q^{n-i+2}-1)\cdots(q^n-1)}{(q-1)(q^2-1)\cdots(q^i-1)} + (-1)^n n = 0$$

where $\beta(1) = \gamma(1) = 1$.

In order to prove that $\beta(n) = q^{1+2+\cdots+(n-1)}$ it is clearly sufficient to prove that for any $n \geq 2$ we have the identity:

$$X = q^{1+2+\cdots+(n-1)} + \sum_{i=1}^{n-1}(-1)^{n-i}q^{1+2+\cdots+(i-1)}$$

$$\times \frac{(q^{n-i+1}-1)(q^{n-i+2}-1)\cdots(q^n-1)}{(q-1)(q^2-1)\cdots(q^i-1)} + (-1)^n = 0 .$$

Now substitute:

$$\frac{(q^{n-i+1}-1)(q^{n-i+2}-1)\cdots(q^n-1)}{(q-1)(q^2-1)\cdots(q^i-1)} = \frac{(q^{n-i+1}-1)(q^{n-i+2}-1)\cdots(q^{n-1}-1)}{(q-1)(q^2-1)\cdots(q^{i-1}-1)}$$

$$+\, q^i\,\frac{(q^{n-i}-1)(q^{n-i+1}-1)\cdots(q^{n-1}-1)}{(q-1)(q^2-1)\cdots(q^i-1)}.$$

We have then

$$X = q^{1+2+\cdots+(n-1)} + \sum_{i=1}^{n-1}(-1)^{n-i}q^{1+2+\cdots+(i-1)}$$

$$\times\frac{(q^{n-i+1}-1)(q^{n-i+2}-1)\cdots(q^{n-1}-1)}{(q-1)(q^2-1)\cdots(q^i-1)}$$

$$+\sum_{i=1}^{n-1}(-1)^{n-i}q^{1+2+\cdots+i}\,\frac{(q^{n-i}-1)(q^{n-i+1}-1)\cdots(q^{n-1}-1)}{(q-1)(q^2-1)\cdots(q^i-1)} + (-1)^n = 0$$

and the formula for $\beta(n)$ is proved. (The formula for $\beta(n)$ is of course, well known, see [14], [15].)

In order to compute $\gamma(n)$, we add together the recurrence formula for $\gamma(n)$ with the corresponding formula in which n is replaced by $n+1$. Note that: $(-1)^n n + (-1)^{n+1}(n+1) = (-1)^{n+1}$ and

$$(-1)^{n-i}\,\frac{(q^{n-i+1}-1)(q^{n-i+2}-1)\cdots(q^n-1)}{(q-1)(q^2-1)\cdots(q^i-1)}$$

$$+\,(-1)^{n-i+1}\,\frac{(q^{n-i+2}-1)(q^{n-i+3}-1)\cdots(q^{n+1}-1)}{(q-1)(q^2-1)\cdots(q^i-1)}$$

$$= (-1)^{n-i+1}\,q^{n-i+1}\,\frac{(q^{n-i+2}-1)(q^{n-i+3}-1)\cdots(q^n-1)}{(q-1)(q^2-1)\cdots(q^{i-1}-1)}.$$

We get:

$$\gamma(n+1) + \sum_{i=1}^{n} (-1)^{n-i+1} \gamma(i) q^{n-i+1} \frac{(q^{n-i+2}-1)(q^{n-i+3}-1)\cdots(q^n-1)}{(q-1)(q^2-1)\cdots(q^{i-1}-1)}$$

$$+ (-1)^{n+1} = 0 \quad (n \geq 2) .$$

This is exactly the recurrence formula satisfied by $\alpha(n)$. Since clearly $\gamma(2) = q-1$ it follows that $\gamma(n+1) = \alpha(n)$ $(n \geq 1)$ and the desired formula for $\gamma(n)$ follows.

We now compute $\delta(n)$. First observe that the number of i-dimensional affine subspaces of V which do not contain 0 equals the number of i-dimensional affine subspaces of V minus the number of i-dimensional linear subspaces of V hence it is given by

$$q^{n-i} \frac{(q^{n-i+1}-1)(q^{n-i+2}-1)\cdots(q^n-1)}{(q-1)(q^2-1)\cdots(q^i-1)} - \frac{(q^{n-i+1}-1)(q^{n-i+2}-1)\cdots(q^n-1)}{(q-1)(q^2-1)\cdots(q^i-1)}$$

$$= \frac{(q^{n-i}-1)(q^{n-i+1}-1)\cdots(q^n-1)}{(q-1)(q^2-1)\cdots(q^i-1)}$$

It follows then from 1.13 (d) that

$$\delta(n) = \sum_{i=0}^{n-1} (-1)^{n-i+1} \alpha(i) \frac{(q^{n-i}-1)(q^{n-i+1}-1)\cdots(q^n-1)}{(q-1)(q^2-1)\cdots(q^i-1)} + (-1)^n$$

$$= \sum_{i=0}^{n-1} (-1)^{n-i-1} (q^{n-i}-1)(q^{n-i+1}-1)\cdots(q^n-1) + (-1)^n$$

which is the desired formula for $\delta(n)$.

Finally we compute $\varepsilon(n,m)$. First observe that the number of (n−i)-dimensional linear subspaces V_{n-i} of V such that $V_{n-i} + \tilde{V} = V$ equals

$$q^{i(n-m)} \frac{(q^{m-i+1}-1)(q^{m-i+2}-1)\cdots(q^m-1)}{(q-1)(q^2-1)\cdots(q^i-1)} .$$

(Note that we must have $i \leq m$.)

It follows from 1.13(e) that for any n, m such that $n > m \geq 1$,

$$\varepsilon(n,m) + \sum_{i=1}^{m} (-1)^i \varepsilon(n-i, m-i) q^{i(n-m)} \frac{(q^{m-i+1}-1)(q^{m-i+2}-1)\cdots(q^m-1)}{(q-1)(q^2-1)\cdots(q^i-1)}$$

$$+ (-1)^{m+1} = 0$$

where $\varepsilon(n-m,0) = 1$. Let $\varepsilon'(n,m) = \varepsilon(n,m)(q-1)(q^2-1)\cdots(q^{n-m-1}-1)$. Then we have

$$\varepsilon'(n,m) + \sum_{i=1}^{m} (-1)^i \varepsilon'(n-i, m-i) q^{i(n-m)} \frac{(q^{m-i+1}-1)(q^{m-i+2}-1)\cdots(q^m-1)}{(q-1)(q^2-1)\cdots(q^i-1)}$$

$$+ (-1)^{m+1}(q-1)(q^2-1)\cdots(q^{n-m-1}-1) = 0 \ .$$

$(n > m \geq 1)$ where $\varepsilon'(n-m,0) = (q-1)(q^2-1)\cdots(q^{n-m-1}-1)$. We wish to prove that $\varepsilon(n,m) = (q^{n-m}-1)(q^{n-m+1}-1)\cdots(q^{n-1}-1)$ or, equivalently, that $\varepsilon'(n,m) = (q-1)(q^2-1)\cdots(q^{n-1}-1)$. It is clearly sufficient to prove that $\varepsilon''(n,m) = (q-1)(q^2-1)\cdots(q^{n-1}-1)$ where $\varepsilon''(n,m)$ is defined by

$$\varepsilon''(n,m) = \sum_{i=1}^{m} (-1)^{i+1}(q-1)(q^2-1)\cdots(q^{n-i-1}-1)q^{i(n-m)}$$

$$\times \frac{(q^{m-i+1}-1)(q^{m-i+2}-1)\cdots(q^m-1)}{(q-1)(q^2-1)\cdots(q^i-1)} + (-1)^m(q-1)(q^2-1)\cdots(q^{n-m-1}-1) \ .$$

We compute the difference $\varepsilon''(n,m+1) - \varepsilon''(n,m)$ where $n > m+1 > m \geq 1$. Observe that

$$q^{i(n-m-1)}(q^{m-i+2}-1)(q^{m-i+3}-1)\cdots(q^{m+1}-1)$$

$$- q^{i(n-m)}(q^{m-i+1}-1)(q^{m-i+2}-1)\cdots(q^m-1)$$

$$= q^{i(n-m-1)}(q^{m-i+2}-1)(q^{m-i+3}-1)\cdots(q^m-1)(q^i-1) \ .$$

We have

$$\varepsilon''(n,m+1) - \varepsilon''(n,m) = \sum_{i=1}^{m+1} (-1)^{i+1}(q-1)(q^2-1)\cdots(q^{n-i-1}-1)$$

$$\times\ q^{i(n-m-1)} \frac{(q^{m-i+2}-1)(q^{m-i+3}-1)\cdots(q^m-1)}{(q-1)(q^2-1)\cdots(q^i-1)}$$

$$+\ (-1)^{m+1}(q-1)(q^2-1)\cdots(q^{n-m-2}-1)q^{n-m-1}$$

$$= q^{n-m-1}((q-1)(q^2-1)\cdots(q^{n-2}-1) - \varepsilon''(n-1,m)) \ .$$

This implies clearly the desired formula by induction on $n+m$. (The formula is obvious for $m = 1$.)

CHAPTER 2

THE AFFINE STEINBERG MODULE

2.1. Let E be an affine space of dimension $\ell \geq 1$ over a finite field F with q elements. The *affine Steinberg module* associated to E is by definition the free A-module $\Delta_A(E)$ defined by using the affine flags in E (see 1.14). Let $\mathrm{Flag}(E)$ be the set of all complete affine flags $\varepsilon = (E_0 \subset E_1 \subset \cdots \subset E_{\ell-1})$ in E ($\dim E_i = i$). Then $\Delta_A(E)$ can be identified with the set of all functions $u : \mathrm{Flag}(E) \to A$, satisfying the "cycle condition"

$$\sum_{\tilde{E}_i} u(E_0 \subset E_1 \subset \cdots \subset E_{i-1} \subset \tilde{E}_i \subset E_{i+1} \subset \cdots \subset E_{\ell-1}) = 0$$

for any given i ($0 \leq i \leq \ell-1$) and given $E_0 \subset E_1 \subset \cdots \subset E_{i-1} \subset E_{i+1} \subset \cdots \subset E_{\ell-1}$ of dimension $0, 1, \cdots, i-1, i+1, \cdots, \ell-1$ (this sum has q terms for $i = 0$ and $q+1$ terms for $0 < i \leq \ell-1$).

Let $\mathrm{Aff}(E)$ be the group of all affine isomorphisms $t : E \xrightarrow{\approx} E$. $\mathrm{Aff}(E)$ acts on $\Delta_A(E)$ by the formula

$$(tu)(E_0 \subset E_1 \subset \cdots \subset E_{\ell-1}) = u(t^{-1}(E_0) \subset t^{-1}(E_1) \subset \cdots \subset t^{-1}(E_{\ell-1}))$$

where $t \in \mathrm{Aff}(E)$ and $u \in \Delta_A(E)$.

Given a flag $\varepsilon = (E_0 \subset E_1 \subset \cdots \subset E_{\ell-1})$ we denote by B_ε the group all $t \in \mathrm{Aff}(E)$ such that $tE_0 = E_0$, $tE_1 = E_1, \cdots, tE_{n-1} = E_{n-1}$. We have the following

THEOREM. *Assume that* A *is a field of characteristic zero. Let* ε *be a complete affine flag in* E. *Then there exists a unique (up to a non-zero scalar) function* $u : \mathrm{Flag}(E) \to A$ *such that*

(i) $u \neq 0$

(ii) $tu = u$ *for all* $t \in B_{\mathcal{E}}$

(iii) $u \in \Delta_A(E)$.

The proof will be given in 2.3.

COROLLARY. *If* A *is a field of characteristic zero,* $\Delta_A(E)$ *is an irreducible* Aff(E)*-module.*

This follows by applying Frobenius duality to the Aff(E)-module $\Delta_A(E)$ and the unit representation of $B_{\mathcal{E}}$.

2.2 *Bruhat decomposition in the affine case*

Let V be an n-dimensional vector space over $F (n \geq 2)$ and let Flag(V) be the set of all complete linear flags $\mathcal{V} = (V_1 \subset V_2 \subset \cdots \subset V_{n-1})$ in $V (\dim V_i = i)$. Given $\mathcal{V} \in$ Flag V, let $B_{\mathcal{V}} = \{t \in GL(V) | tV_1 = V_1, tV_2 = V_2, \cdots, tV_{n-1} = V_{n-1}\}$. The group GL(V) acts transitively on Flag(V) by the formula $t(V_1 \subset V_2 \subset \cdots \subset V_{n-1}) = (tV_1 \subset tV_2 \subset \cdots \subset tV_{n-1})$. This action is not transitive when restricted to $B_{\mathcal{V}}$. The orbits of $B_{\mathcal{V}}$ on Flag(V) are described by Bruhat's theorem. Let us recall the content of this theorem. Given $\mathcal{V}' = (V_1' \subset V_2' \subset \cdots \subset V_{n-1}') \in$ Flag(V) we define a permutation $b(\mathcal{V}') = (i_1, i_2, \cdots, i_n)$ of $(1, 2, \cdots, n)$ by the formulae: $V_1' \cap V_{i_1-1} \neq V_1' \cap V_{i_1}$, $V_2' \cap V_{i_2-1} \neq V_2' \cap V_{i_2}$ $(i_2 \neq i_1), \cdots, V_n' \cap V_{i_n-1} \neq V_n' \cap V_{i_n}$ $(i_n \neq i_1, i_2, \cdots, i_{n-1})$. Here we use the convention: $V_0 = 0$, $V_n = V_n' = V$. Bruhat's theorem states that $\mathcal{V}', \mathcal{V}''$ are in the same $B_{\mathcal{V}}$-orbit if and only if $b(\mathcal{V}') = b(\mathcal{V}'')$ and that the map $b : \{B_{\mathcal{V}}\text{-orbits}\} \to \{\text{permutations of } 1, 2, \cdots, n\}$ is bijective.

We need an extension of this result to the affine case. Such an extension has been found by Solomon [12] but we shall need a somewhat different approach. The affine space E can be regarded as an affine hyperplane not containing the origin in an $(\ell+1)$-dimensional vector space V. Let H be the unique linear hyperplane in V which is parallel to E. Note that

Aff(E) can be regarded as the set of all $t \in GL(V)$ such that $t(E) = E$. Given $\varepsilon = (E_0 \subset E_1 \subset \cdots \subset E_{\ell-1})$ our problem is to classify the orbits of B_ε on Flag(E). (Note that Aff(E) acts transitively on Flag(E).)

ε gives rise to a complete linear flag $[\varepsilon] = ([E_0] \subset [E_1] \subset \cdots \subset [E_{\ell-1}])$ in V and to a complete linear flag $[\varepsilon]_H = ([E_1] \cap H \subset [E_2] \cap H \subset \cdots \subset [E_{\ell-1}] \cap H)$ in H. (Here $[E_i]$ is the linear span of E_i in V.) Let $\varepsilon' = (E'_0 \subset E'_1 \subset \cdots \subset E'_{\ell-1}) \in$ Flag(E). ε' gives rise similarly to complete linear flags $[\varepsilon']$, $[\varepsilon']_H$ in V and H. Define a permutation $(i_0, i_1, \cdots, i_\ell)$ of $(0, 1, \cdots, \ell)$ by the formulae

$$[E'_0] \cap [E_{i_0}] \neq [E'_0] \cap [E_{i_0-1}],$$

$$[E'_1] \cap [E_{i_1}] \neq [E'_1] \cap [E_{i_1-1}] \ (i_1 \neq i_0), \cdots,$$

$$[E'_\ell] \cap [E_{i_\ell}] \neq [E'_\ell] \cap [E_{i_\ell-1}], \ (i_\ell \neq i_0, i_1, \cdots, i_{\ell-1}).$$

Here we use the convention $[E_{-1}] = 0$, $[E_\ell] = [E'_\ell] = V$. Similarly define a permutation $(j_1, j_2, \cdots, j_\ell)$ of $(1, 2, \ldots, \ell)$ by the formulae

$$[E'_1] \cap [E_{j_1}] \cap H \neq [E'_1] \cap [E_{j_1-1}] \cap H,$$

$$[E'_2] \cap [E_{j_2}] \cap H \neq [E'_2] \cap [E_{j_2-1}] \cap H (j_2 \neq j_1), \cdots,$$

$$[E'_\ell] \cap [E_{j_\ell}] \cap H \neq [E'_\ell] \cap [E_{j_\ell-1}] \cap H (j_\ell \neq j_1, j_2, \cdots, j_{\ell-1}).$$

Let $\tilde{b}(\varepsilon') = \binom{i_0, i_1, i_2, \cdots, i_\ell}{j_1, j_2, \cdots, j_\ell}$; this is a pair of permutations: one of $0, 1, \cdots, \ell$ and one of $1, 2, \cdots, \ell$.

This invariant depends clearly only on the B_ε-orbit of ε', and it is not difficult to see that it distinguishes B_ε-orbits (i.e., $\tilde{b}(\varepsilon') = \tilde{b}(\varepsilon'')$ $\Longleftrightarrow$ $\varepsilon', \varepsilon''$ are in the same B_ε-orbit). Note however that the invariant $\tilde{b}(\varepsilon')$ cannot take arbitrary values. In fact, the entries of $\tilde{b}(\varepsilon')$ must satisfy

$$j_1 \in \{i_0, i_1\}, \ j_2 \in \{i_0, i_1, i_2\}, \cdots, j_\ell \in \{i_0, i_1, i_2, \cdots, i_\ell\} \tag{3}$$

$$j_1 \geq i_1, \ j_2 \geq i_2, \cdots, j_\ell \geq i_\ell . \tag{4}$$

To prove this, let

$$d_a = \dim([E'_a] \cap [E_{j_a}] \cap H)/[E'_a] \cap [E_{j_a - 1}] \cap H$$

$$d'_a = \dim([E'_a] \cap [E_{j_a}])/[E'_a] \cap [E_{j_a - 1}]. \quad (1 \leq a \leq \ell)$$

It is clear that $0 \leq d_a \leq d'_a \leq 1$. Hence $d_a \neq 0 \Longrightarrow d'_a \neq 0$ which clearly implies (3).

On the other hand, for any a, $1 \leq a \leq \ell$, let

$$\delta_a = \dim([E'_a] \cap [E_{j_a}]) = \text{number of indices } i_0, i_1, \cdots, i_a$$

which are $\leq ja$.

$$\delta'_a = \dim([E'_a] \cap [E_{j_a}] \cap H) = \text{number of indices } j_1, j_2, \cdots, j_a$$

which are $\leq ja$.

$$\delta''_a = \dim([E'_{a-1}] \cap [E_{j_a}]) = \text{number of indices } i_0, i_1, \cdots, i_{a-1}$$

which are $\leq ja$.

$$\delta'''_a = \dim([E'_{a-1}] \cap [E_{j_a}] \cap H) = \text{number of indices } j_1, j_2, \cdots, j_{a-1}$$

which are $\leq ja$. Assume that $j_a < i_a$. It follows that $\delta_a = \delta'_a = \delta''_a$ and $\delta'''_a = \delta_a - 1$. Hence $[E'_a] \cap [E_{j_a}] = [E'_a] \cap [E_{j_a}] \cap H = [E'_{a-1}] \cap [E_{j_a}]$. It follows that

$$[E'_{a-1}] \cap [E_{j_a}] \cap H = ([E'_{a-1}] \cap [E_{j_a}]) \cap ([E'_a] \cap [E_{j_a}] \cap H)$$
$$= [E'_a] \cap [E_{j_a}]$$

which shows that $\delta'''_a = \delta_a$, a contradiction. This proves (4).

We have the following

PROPOSITION. *The invariant* b *defines a* 1-1 *correspondence between the set of* B_ε*-orbits in* Flag(E) *and the set of arrays* $\binom{i_0\ i_1\ i_2 \cdots i_\ell}{\ \ j_1\ j_2 \cdots j_\ell}$ *as above, satisfying conditions* (3) *and* (4).

The proof will be left to the reader. (See the paper [12] of Solomon where he describes a 1-1 correspondence between the B_ε-orbits in Flag(E) and some set of invariants $(j_1, j_2, \cdots, j_\ell; \gamma_1, \gamma_2, \cdots \gamma_\ell)$ where $j_1, j_2, \cdots, j_\ell$ are as above and $\gamma_1, \gamma_2, \cdots, \gamma_\ell$ are some integers modulo 2.)

COROLLARY. *Let* $\varepsilon' = (E'_0 \subset E'_1 \subset \cdots \subset E'_{\ell-1})$ *and* $\varepsilon'' = (E''_0 \subset E''_1 \subset \cdots \subset E''_{\ell-1})$ *in* Flag(E). *Then* $\varepsilon', \varepsilon''$ *are in the same* B_ε*-orbit if and only if* $\dim([E'_i] \cap [E_j]) = \dim([E''_i] \cap [E_j])$ *for all* $i, j \epsilon \{0, 1, \cdots, \ell-1\}$ *and* $\dim([E'_i] \cap [E_j] \cap H) = \dim([E''_i] \cap [E_j] \cap H)$ *for all* $i, j \epsilon \{1, 2, \cdots, \ell-1\}$.

2.3 We are now ready for the

Proof of Theorem 2.1. We say that a flag $\varepsilon' = (E'_0 \subset E'_1 \subset \cdots \subset E'_{\ell-1})$ in E is of type I if for any $i, j \epsilon \{0, 1, \cdots, \ell-1\}$ we have the implication $[E'_i] \cap [E_j] \neq 0 \Longrightarrow [E'_i] \cap [E_j] \not\subset H$. Otherwise we say that ε' is of type II. Note that ε' is of type I if and only if its invariant $\tilde{b}(\varepsilon')$ has the property: $j_a = \max(i_a, \min(i_0, i_1, \cdots, i_{a-1}))$ for all $a = 1, 2, \cdots, \ell$. (Hence the sequence $(j_1, j_2, \cdots, j_\ell)$ is completely determined by the sequence $(i_0, i_1, \cdots, i_\ell)$.)

We first prove

LEMMA. *Let* u *be a function as in the Theorem* 2.1. *Then* u *must vanish on all flags of type* II.

In fact, let $\varepsilon' \epsilon$ Flag(E) be of type II. Let $a = a(\varepsilon')$ be the smallest integer such that there exists an integer b with $[E'_a] \cap [E_b] \neq 0$, $[E'_a] \cap [E_b] \subset H$. We clearly have $a \geq 1$. We shall prove the lemma by induction on a. We can choose b to be minimal with the above property.

Then we have: $[E'_a] \cap [E_b] = L$ (line), $L \subset H$, $[E'_a] \cap [E_{b-1}] = 0$ and $[E'_{a-1}] \cap [E_b] = 0$. Keeping fixed $[E'_0], [E'_1], \cdots, [E'_{a-2}], [E'_a], \cdots, [E'_{\ell-1}]$ we consider all (a−1)-dimensional affine subspaces $\tilde{E}_{a-1}$ of E which lie between E'_{a-2} and E'_a. (There are (q+1) such subspaces if $a > 1$ and q if $a = 1$.) There are exactly q values of $\tilde{E}_{a-1}$ such that $[\tilde{E}_{a-1}] \cup [E_b] = 0$; moreover the flags $(E'_0 \subset E'_1 \subset \cdots \subset E'_{a-2} \subset \tilde{E}_{a-1} \subset E'_a \subset \cdots \subset E'_{\ell-1})$ corresponding to these values of $\tilde{E}_{a-1}$ are all in the same B_ε-orbit (cf. Corollary 2.2), and they include the original flag ε'. In the case $a > 1$ there is a unique value for $\tilde{E}_{a-1}$ such that $[\tilde{E}_{a-1}] \cap [E_b] \neq 0$ this is precisely $E'_{a-2} \oplus L$. Using the cycle condition for u and the fact that u is B_ε-invariant we get:

$$q \cdot u(\varepsilon') = \begin{cases} -u(\varepsilon'') & \text{if } a > 1 \\ 0 & \text{if } a = 1 \end{cases}$$

where $\varepsilon'' = (E'_0 \subset E'_1 \subset \cdots \subset E'_{a-2} \subset E'_{a-2} \oplus L \subset E'_a \subset \cdots \subset E'_{\ell-1})$. Hence $a(\varepsilon'') = a(\varepsilon') - 1$. From this the lemma follows by induction on a. (Note that q is not a zero divisor in Λ.)

We now concentrate on flags of type I. Note that these are in 1-1 correspondence with the permutations $(i_0, i_1, \cdots, i_\ell)$ of $(0, 1, \cdots, \ell)$, since the invariants $j_1, \cdots, j_\ell$ are uniquely determined. Let u be a function as in the Theorem 2.1. Since u is constant on B_ε-orbits and since it vanishes on flags of type II (by the lemma) we can regard u as a map $u: \{\text{permutations of } 0, 1, \cdots, \ell\} \to \Lambda$.

Let $(E'_0 \subset E'_1 \subset \cdots \subset E'_{a-2} \subset E'_a \subset \cdots \subset E'_{\ell-1})$ be a subcomplete flag in E (the (a−1)-dimensional subspace is missing). Associate to this a permutation $(i_0, i_1, \cdots, i_\ell)$ of $(0, 1, \cdots, \ell)$ with $i_{a-1} > i_a$ by the formulae:

$$[E'_0] \cap [E_{i_0}] \neq [E'_0] \cap [E_{i_0-1}], \cdots, [E'_{a-2}] \cap [E_{i_{a-2}}] \neq [E'_{a-2}] \cap [E_{i_{a-2}-1}]$$

$$[E'_a] \cap [E_{i_{a-1}}] \neq [E'_a] \cap [E_{i_{a-1}-1}],\ [E'_a] \cap [E_{i_a}] \neq [E'_a] \cap [E_{i_a-1}],$$

$$[E'_{a+1}] \cap [E_{i_{a+1}}] \neq [E'_{a+1}] \cap [E_{i_{a+1}-1}], \cdots, [E'_\ell] \cap [E_{i_\ell}] \neq [E'_\ell] \cap [E_{i_\ell-1}].$$

(Convention: $[E_{-1}] = 0$, $[E_\ell] = [E'_\ell] = V$.) We assume that $[E'_i] \cap [E_j] \neq 0 \Longrightarrow [E'_i] \cap [E_j] \not\subset H$ for all $i \neq a-1$ and all j. We shall make $(E'_0 \subset E'_1 \subset \cdots \subset E'_{a-2} \subset E'_a \subset \cdots \subset E'_{\ell-1})$ into a complete flag by filling in with the missing E'_{a-1}'s in all possible ways.

Case 1. $(a = 1)$. There are q possibilities for E'_0. One of them is $E'_0 = E'_1 \cap E_{i_1}$. The remaining $(q-1)$ possibilities give rise to complete flags in the same B_ε-orbit (cf. Corollary 2.2). In all q cases we get complete flags of type I with invariant $(i_1, i_0, i_2, i_3, \cdots, i_\ell)$ for $E'_1 \cap E_{i_1}$ and $(i_0, i_1, \cdots, i_\ell)$ for the remaining ones. From the cycle condition it follows that $(q-1)\, u(i_0, i_1, \cdots, i_\ell) + u(i_1, i_0, i_2, \cdots, i_\ell) = 0$, $(i_1 > i_2)$. Next assume $a \geq 2$; there are now $(q+1)$ possibilities for E'_{a-1}.

Case 2. $i_{a-1} \neq \min(i_0, i_1, \cdots, i_{a-1})$ $(a \geq 2)$. One possibility is $E'_{a-1} = \langle E'_{a-2}, E'_a \cap E_{i_a} \rangle$ (affine span). The remaining q values of E'_{a-1} give rise to complete flags in the same B_ε-orbit (cf. Corollary 2.2). In all $(q+1)$ cases we get complete flags of type I with invariant $(i_0, i_1, \cdots, i_{a-2}, i_a, i_{a-1}, i_{a+1}, \cdots, i_\ell)$ for the first value of E'_{a-1} and $(i_0, i_1, \cdots, i_{a-2}, i_{a-1}, i_a, \cdots, i_\ell)$ for the remaining q values. From the cycle condition it follows that $q\,u(i_0, i_1, \cdots, i_{a-2}, i_{a-1}, i_a, \cdots, i_\ell) + u(i_0, i_1, \cdots, i_{a-2}, i_a, i_{a-1}, i_{a+1}, \cdots, i_\ell) = 0$.

Case 3. $i_{a-1} = \min(i_0, i_1, \cdots, i_{a-1})$ $(a \geq 2)$. In this case $L = [E'_a] \cap [E_{i_{a-1}}] \cap H$ is a line not contained in $[E'_{a-2}]$, so we can take $E'_{a-1} = ([E'_{a-2}] + L) \cap E$ and this gives rise to a complete flag of type II (on which u vanishes by the lemma). Another possibility is $E'_{a-1} = \langle E'_{a-2}, E'_a \cap E_{i_a} \rangle$ (affine span). This gives rise to a complete flag of type I with invariant $(i_0, i_1, \cdots, i_{a-2}, i_a, i_{a-1}, i_{a+1}, \cdots, i_\ell)$. The remaining $(q-1)$ possibilities for E'_{a-1} give rise to complete flags of type I, in the same B_ε-orbit (cf. Corollary 2.2) with invariant $(i_0, i_1, \cdots, i_{a-2}, i_{a-1}, i_a, i_{a+1}, \cdots, i_\ell)$. From the cycle condition it follows that

$$(q-1)\, u(i_0, i_1, \cdots, i_{a-2}, i_{a-1}, i_a, i_{a+1}, \cdots, i_\ell) + u(i_0, i_1, \cdots, i_{a-2}, i_a, i_{a-1}, i_{a+1}, \cdots, i_\ell) = 0\ .$$

We can collect the three cases in a single formula:

$$\theta \cdot u(i_0, i_1, \cdots, i_{a-2}, i_{a-1}, i_a, i_{a+1}, \cdots, i_\ell) + u(i_0, i_1, \cdots, i_{a-2}, i_a, i_{a-1}, i_{a+1}, \cdots, i_\ell) = 0 \tag{5}$$

where

$$\theta = \begin{cases} q & \text{if } i_{a-1} \neq \min(i_0, i_2, \cdots, i_{a-1}) \\ q-1 & \text{otherwise.} \end{cases}$$

Given a permutation $w = (i_0, i_1, \cdots, i_\ell)$ of $(0, 1, \cdots, \ell)$ we define its length $\ell(w)$, as usual, as the minimal number of fundamental transpositions $(a, a+1)$, $0 \leq a \leq \ell-1$ of which w can be the product.

We also define $m(w)$ as the number of indices $i_a (0 \leq a \leq \ell-1)$ such that $i_a \neq 0$ and $i_a = \min(i_0, i_1, \cdots, i_a)$. It is clear that always $\ell(w) \geq m(w)$. Let $w_0 = (\ell, \ell-1, \cdots, 1, 0)$ be the unique permutation of maximal length $\ell(w_0) = 1 + 2 + \cdots + \ell$. It is clear that $m(w_0) = \ell$. Let $w'_0 = (0, 1, \cdots, \ell-1, \ell)$; we have $\ell(w'_0) = 0$, $m(w'_0) = 0$. From (5) we get immediately by induction on the length that

$$u(w) = (-q)^{-\ell(w)+m(w)}(1-q)^{-m(w)} u(w'_0) .$$

It follows that $u(w) = q^{\ell(w_0)-\ell(w)+m(w)-m(w_0)}(q-1)^{m(w_0)-m(w)} u(w_0)$ where w is any permutation of $(0, 1, \cdots, \ell)$. (It is easy to see that in the last formula all exponents are positive.) This proves the unicity of u up to a scalar. The same proof shows the existence of u. The theorem is proved.

Remark. Define a function $u' : \{\text{complete flags of type I in } E\} \to \mathbf{Z}$ by the formula $u'(\varepsilon') =$ number of elements in the B_ε-orbit of ε'. u' satisfies a recurrence formula similar to the one satisfied by u:

$$u'(i_0, i_1, \cdots, i_{a-2}, i_{a-1}, i_a, i_{a+1}, \cdots, i_\ell) = \theta \cdot u'(i_0, i_1, \cdots, i_{a-2}, i_a, i_{a-1}, i_{a+1}, \cdots, i_\ell) ,$$

where θ is as in (5) and B_ε-orbits of type I in Flag(E) are identified with permutations of $0,1,\ldots,\ell$. It follows that u and u′ are related by $u(w)\cdot u'(w) = (-1)^{\ell(ww_0)} \times$ constant. Choosing the constant to be equal to the order of B_ε we finally get the following formula for the function u of Theorem 2.1: $u(\varepsilon') = (-1)^{\ell(ww_0)}$ times the number of elements $t \in B_\varepsilon$ such that $t\varepsilon' = \varepsilon'$, if ε' has type I and $u(\varepsilon') = 0$ if ε' has type II.

2.4 *Affine foliations*

Let E be, as above, an ℓ-dimensional affine space over F. Assume that $\ell \geq 2$. An affine foliation Φ of E is by definition a decomposition of E into affine subspaces (called leaves) of fixed dimension $m(0<m<\ell)$ in such a way that any point in E is contained in a unique leaf and any two leaves can be obtained one from the other by a translation in E. If we regard E as an affine hyperplane in an $(\ell+1)$-dimensional vector space V $(0 \notin E)$, giving a foliation of E is equivalent to giving an m-dimensional linear subspace $\tilde{V}$ of V, parallel to E (the leaves of the foliation being those affine subspaces of E which can be obtained from $\tilde{V}$ by a translation in V). The partially ordered set $S_{IV}(V,\tilde{V})$ defined in 1.8 is then canonically isomorphic to the set $S_{IV}(E,\Phi)$ of all affine subspaces E′ of E such that $E' \neq E$ and E′ is *transversal* to the leaves, i.e., E′ meets all leaves of Φ and E is spanned by E′ and any leaf of Φ. The 1-1 correspondence $S_{IV}(V,\tilde{V}) \longleftrightarrow S_{IV}(E,\Phi)$ is given by $V' \longleftrightarrow E' = V' \cap E$. $S_{IV}(E,\Phi)$ is a subset of $S_I(E)$ (see 1.8) and has the induced partial order.

Consider now the affine space E/Φ whose points are the leaves of Φ and whose i-dimensional affine subspaces are the (m+i)-dimensional affine subspaces of E which contain some leaf of Φ. Note that $\dim E/\Phi = \ell - m$. The set $S_I(E/\Phi)$ can be identified with the subset of $S_I(E)$ consisting of all affine subspaces in $S_I(E)$ which contain some leaf of Φ and are different from E. In other words, given Φ, we have two natural subsets $S_{IV}(E,\Phi)$ and $S_I(E/\Phi)$ of $S_I(E)$ with induced partial orders. We define a new partially ordered set $\tilde{S} = \tilde{S}(E,\Phi)$ as follows: as a set we have $\tilde{S} = S_I(E/\Phi) \cup S_{IV}(E,\Phi)$ (disjoint union); the partial order on $\tilde{S}$ is defined by the requirements that any element in $S_I(E/\Phi)$ is $<$ than any element in $S_{IV}(E,\Phi)$ and that the induced partial orders on the subsets $S_I(E/\Phi)$ and $S_{IV}(E,\Phi)$ should be the original ones. It follows immediately that the associated simplicial complex (see 1.3) $X(\tilde{S})$ is precisely the *join* of the simplicial complexes $X(S_I(E/\Phi))$ and $X(S_{IV}(E,\Phi))$. It follows then from the Künneth formula that there is a natural isomorphism

$$\tilde{H}_{\ell-m-1}(S_I(E/\Phi);A) \otimes \tilde{H}_{m-1}(S_{IV}(E,\Phi);A) \approx H_{\ell-1}(\tilde{S};A)\ . \tag{6}$$

(There is no error term in the Künneth formula since the spaces concerned are torsion free, cf. 1.14; $\tilde{H}(\cdot)$ denotes reduced homology, i.e., $\tilde{H}_i(\cdot) = H_i(\cdot)$ if $i > 0$ and $\tilde{H}_0(\cdot) = \ker(H_0(\cdot) \to A)$.)

We now compare $S_I(E)$ and $\tilde{S}(E,\Phi)$. Define a map $\alpha : S_I(E) \to \tilde{S}$ by the formula

$$\alpha(E') = \begin{cases} E' & \text{if } E' \in S_{IV}(E,\Phi) \\ <E',\mathfrak{L}> & \text{if } E' \in S_I(E) \setminus S_{IV}(E,\Phi)\ . \end{cases}$$

Here $\mathfrak{L}$ is any leaf of Φ meeting E'.

The map α respects the partial order. In fact, let $E', E'' \in S_I(E)$ such that $E' \subset E''$. Assume first that $E', E'' \in S_{IV}(E,\Phi)$. Then $\alpha(E') = E' \subset E'' = \alpha(E'')$ hence $\alpha(E') \leq \alpha(E'')$. Assume next that $E', E'' \in S_I(E) \setminus S_{IV}(E,\Phi)$. Let $\mathfrak{L}$ be a leaf meeting E'. We have $\alpha(E') = <E',\mathfrak{L}> \subset <E'',\mathfrak{L}> = \alpha(E'')$, hence $\alpha(E') \leq \alpha(E'')$. Finally, assume that $E' \in S_I(E) \setminus S_{IV}(E,\Phi)$, $E'' \in S_{IV}(E,\Phi)$. Then clearly $\alpha(E') \in S_I(E/\Phi)$ hence $\alpha(E') < \alpha(E'') = E''$ by the definition of the partial order in $\tilde{S}(E,\Phi)$. Note that the case $E' \in S_{IV}(E,\Phi)$, $E'' \in S_I(E) \setminus S_{IV}(E,\Phi)$ cannot occur when $E' \subset E''$. This proves that α respects the partial order.

It follows that α induces a simplicial map $X(S_I(E)) \to X(\tilde{S})$ and hence also a map $\tilde{\alpha}_A : H_{\ell-1}(S_I(E);A) \to H_{\ell-1}(\tilde{S};A)$. We shall prove that

$\tilde{\alpha}_A$ is an isomorphism. Since $\tilde{\alpha}_A = \tilde{\alpha}_Z \otimes 1_A$ it is sufficient to prove that $\tilde{\alpha}_Z$ is an isomorphism, and this would follow if we can prove that $\tilde{\alpha}_A$ is an isomorphism whenever A is a field. Assume now that A is a field. We know that $\dim_A H_{\ell-1}(S_I(E); A) = (q-1)(q^2-1)\cdots(q^\ell-1)$ (cf. 1.14) and $\dim_A H_{\ell-1}(\tilde{S}; A) = \dim_A \tilde{H}_{\ell-m-1}(S_I(E/\Phi); A)$. $\dim_A \tilde{H}_{m-1}(S_{IV}(E,\Phi); A) = (q-1)(q^2-1)\cdots(q^{\ell-m}-1)\cdot(q^{\ell-m+1}-1)(q^{\ell-m+2}-1)\cdots(q^\ell-1)$ (cf. 1.14). It follows that $\tilde{\alpha}_A$ is a map between A-vector spaces of equal dimension, hence it is sufficient to prove that $\tilde{\alpha}_A$ is injective. To see this we first observe that given any $(\ell-1)$ simplex $\sigma = (E_m \subset E_{m+1} \subset \cdots \subset E_{\ell-1} \leq E'_{\ell-m} \subset E'_{\ell-m+1} \subset \cdots \subset E'_{\ell-1})$ in $X(\tilde{S})$ (where E_i contains some leaf, $m \leq i \leq \ell-1$ and E'_i is transversal to leaves, $\ell-m \leq i \leq \ell-1$) there is a unique $(\ell-1)$ simplex $\tau = (\tilde{E}_0 \subset \tilde{E}_1 \subset \cdots \subset \tilde{E}_{\ell-1})$ in $X(S_I(E))$ which under α maps isomorphically onto σ. In fact we have $\tilde{E}_i = E_{m+i} \cap E'_{\ell-m}$ $(0 \leq i \leq \ell-m-1)$ and $E_i = E'_i (\ell-m \leq i \leq \ell-1)$. All other $(\ell-1)$ simplices of $X(S_I(E))$ are mapped by α onto simplices of lower dimension. It follows that given a chain $u \in C_{\ell-1}(X(S_I(E)); A)$, the image $\tilde{\alpha}_A u \in C_{\ell-1}(X(\tilde{S}); A)$ is given by the formula:

$$(\tilde{\alpha}_A u)(E_m \subset E_{m+1} \subset \cdots \subset E_{\ell-1} \leq E'_{\ell-m} \subset E'_{\ell-m+1} \subset \cdots \subset E'_{\ell-1})$$

$$= u(E_m \cap E'_{\ell-m} \subset E_{m+1} \cap E'_{\ell-m} \subset \cdots \subset E_{\ell-1} \cap E'_{\ell-m} \subset E'_{\ell-m} \subset E'_{\ell-m+1} \subset \cdots \subset E'_{\ell-1}$$

This shows that $\alpha_A u = 0$ if and only if u vanishes on all complete flags $(\tilde{E}_0 \subset \tilde{E}_1 \subset \cdots \subset \tilde{E}_{\ell-1})$ in E such that $\dim(\tilde{E}_i \cap \mathcal{L}) = \max(0, i+m-\ell)$, $0 \leq i \leq \ell-1$ ($\mathcal{L}$ is the leaf through $\tilde{E}_0$). Note that these are precisely the flags in general position with respect to the leaves of the foliation.

We have the following

LEMMA. *Let* $u \in C_{\ell-1}(X(S_I(E); A)$ *be a cycle. If* u *vanishes on all complete flags* $(\tilde{E}_0 \subset \tilde{E}_1 \subset \cdots \subset \tilde{E}_{\ell-1})$ *such that* $\dim(\tilde{E}_i \cap \mathcal{L}) = \max(0, i+m-\ell)$, $0 \leq i \leq \ell-1$, *then* $u = 0$. *Hence the map* $\tilde{\alpha}_A : H_{\ell-1}(X(S_I(E)); A) \to H_{\ell-1}(X(\tilde{S}); A))$ *is injective.*

Proof. For any complete flag $\tilde{\varepsilon} = (\tilde{E}_0 \subset \tilde{E}_1 \subset \cdots \subset \tilde{E}_{\ell-1})$ in E we define a sequence of numbers $d(\tilde{\varepsilon}) = (d_0, d_1, \cdots, d_{\ell-1}, d_\ell)$ by the formula $d_i = \dim(\tilde{E}_i \cap \mathcal{L})$ $0 \le i \le \ell-1$, $d_\ell = m$. (Recall that $\mathcal{L}$ is the leaf through $\tilde{E}_0$). We have $d_0 = 0$, $d_i \le d_{i+1} \le d_i + 1$ $(0 \le i \le \ell-1)$. We say that $d(\tilde{\varepsilon})$ is the type of $\tilde{\varepsilon}$. We define the level of $\tilde{\varepsilon}$ by the formula $\lambda(\tilde{\varepsilon}) = d_0 + d_1 + \cdots + d_{\ell-1} + d_\ell$, where $d(\tilde{\varepsilon}) = (d_0, d_1, \cdots, d_\ell)$. Note that $\lambda(\tilde{\varepsilon})$ is minimal if and only if $\lambda(\tilde{\varepsilon}) = 1 + 2 + \cdots + m$ or if and only if $d(\tilde{\varepsilon}) = (0, 0, \cdots, 0, 1, 2, 3, \cdots, m)$. We shall prove the lemma by induction on the level of $\tilde{\varepsilon}$. We know that $u(\tilde{\varepsilon}) = 0$ when $\lambda(\tilde{\varepsilon})$ is minimal. Assume now that $\tilde{\varepsilon}$ has non-minimal level. Then there exists some $i, 0 < i \le \ell-1$ such that $d_{i-1} < d_i = d_{i+1}$. Consider all i-dimensional subspaces E_i of $\tilde{E}_{i+1}$ such that $\tilde{E}_{i-1} \subset E_i$. We have

$$\sum_{E_i} u(\tilde{E}_1 \subset \cdots \subset \tilde{E}_{i-1} \subset E_i \subset \tilde{E}_{i+1} \subset \cdots \subset \tilde{E}_{\ell-1}) = 0 .$$

In the above sum all flags except for the original flag $\tilde{\varepsilon}$, have type $(d_0, d_1, \cdots, d_{i-1}, d_i - 1, d_{i+1}, \cdots, d_\ell)$ hence have level equal to $\lambda(\tilde{\varepsilon}) - 1$, so we can assume, by the induction hypothesis that u vanishes on them. It follows that $u(\varepsilon) = 0$ and the lemma is proved.

It follows from the lemma that $\tilde{a}_A$ is actually an isomorphism. Combining this with the isomorphism (6) and identifying $S_{IV}(E, \Phi) = S_{IV}(V, \tilde{V})$, we have the following

THEOREM. *Let* V *be an* $(\ell+1)$*-dimensional vector space over* F, $\tilde{V}$ *an* m*-dimensional linear subspace of* $V (\ell > m > 0)$ *and* E *an affine hyperplane in* V *such that* $0 \notin E$ *and* $\tilde{V}$ *is parallel to* E. *Let* $E/\tilde{V}$ *be the image of* E *under the canonical projection* $V \overset{\pi}{\to} V/\tilde{V}$. *Then* $E/\tilde{V}$ *is an affine hyperplane in* $V/\tilde{V}$ *not containing zero and there exists a canonical isomorphism*

$$\psi_E : \Delta_A(E/\tilde{V}) \otimes P_A(V, \tilde{V}) \overset{\approx}{\to} \Delta_A(E)$$

uniquely characterized by the property that

$$\psi_E(u \otimes u')(E_0 \subset E_1 \subset \cdots \subset E_{\ell-1}) = u(\pi(E_0) \subset \pi(E_1) \subset \cdots \subset \pi(E_{\ell-m-1}))$$

$$\cdot\ u'([E_{\ell-m}] \subset [E_{\ell-m+1}] \subset \cdots \subset [E_{\ell-1}])$$

for all $u \in \Delta_A(E/\tilde{V})$, $u' \in P_A(V, \tilde{V})$ *and all complete affine flags* $(E_0 \subset E_1 \subset \cdots \subset E_{\ell-1})$ *in* E *such that*, $\dim([E_i] \cap \tilde{V}) = \max(0, i+m-\ell)$ $(0 \leq i \leq \ell-1)$.

CHAPTER 3

THE DISTINGUISHED DISCRETE SERIES MODULE

3.1 Let V be a vector space of dimension $n (n \geq 2)$ over a finite field F with q elements. Let Y be the set of all complete affine flags $\varepsilon = (E_0 \subset E_1 \subset \cdots \subset E_{n-1})$ in $V(\dim E_i = i)$ which are away from 0, i.e., $0 \notin E_i (o \leq i \leq n-1)$.

Let $\mathcal{F}_A = \mathcal{F}_A(V)$ be the set of functions $f: Y \to A$. $\mathcal{F}_A$ can be clearly identified with the set of chains $C_{n-1}(X(S_{III}(V)); A)$ (see 1.8). The homology $\dot{\Delta}_A(V) = H_{n-1}(S_{III}(V); A)$ can then be identified with the subspacc of $\mathcal{F}_A$ consisting of all functions $f \in \mathcal{F}_A$ satisfying the "cycle conditions"

$$(7_i) \qquad \sum_{\tilde{E}_i} f(E_0 \subset E_1 \subset \cdots \subset E_{i-1} \subset \tilde{E}_i \subset E_{i+1} \subset \cdots \subset E_{n-1}) = 0$$

for any given i, $0 \leq i \leq n-1$ and given $E_0 \subset E_1 \subset \cdots \subset E_{i-1} \subset E_{i+1} \subset \cdots \subset E_{n-1}$ of dimension $0, 1, \cdots, i-1, i+1, \cdots, n-1$ (this sum has q terms for $i = 0$, $i = n-1$ and $q+1$ terms for $0 < i < n-1$).

We shall now define a natural endomorphism of $\dot{\Delta}_A(V)$. In order to do so we define, for every $\varepsilon \in Y$ a subset $\Theta(\varepsilon) \subset Y$ as follows: $\Theta(\varepsilon)$ is the set of all complete affine flags $(E'_0 \subset E'_1 \subset \cdots \subset E'_{n-1})$ in V such that $E'_0 \in E_{n-1} \setminus E_{n-2}$, $E'_1 \| [E_0], E'_2 \| [E_1], \cdots, E'_{n-1} \| [E_{n-2}]$. Note that all flags in $\Theta(\varepsilon)$ lie in $Y (0 \notin E'_i)$ and that the set $\Theta(\varepsilon)$ has $q^{n-1} - q^{n-1} - q^{n-2}$ elements; in fact a flag in $\Theta(\varepsilon)$ is completely determined by its 0-dimensional member. Define $T: \mathcal{F}_A \to \mathcal{F}_A$ by the formula

$$(Tf)(E_0 \subset E_1 \subset \cdots \subset E_{n-1}) = (-1)^{n-1} \sum f(E'_0 \subset E'_1 \subset \cdots \subset E'_{n-1})$$

where the sum is over all $(E'_0 \subset E'_1 \subset \cdots \subset E'_{n-1}) \in \Theta(E_0 \subset E_1 \subset \cdots \subset E_{n-1})$.

LEMMA. *Suppose that* $f \in \mathcal{F}_A$ *satisfies the cycle conditions. Then* Tf *must also satisfy the cycle conditions.*

Proof. By assumption, f satisfies the cycle conditions (7_i) for $0 \le i \le n-1$. We first prove that Tf satisfies (7_0). Let $(E_1 \subset E_2 \subset \cdots \subset E_{n-1})$ be an incomplete flag in V $(0 \notin E_i)$. We have

$$\sum_{\tilde{E}_0 \neq 0}^{(q)} (Tf)(\tilde{E}_0 \subset E_1 \subset \cdots \subset E_{n-1}) =$$

$$= (-1)^{n-1} \sum_{\substack{(E_0' \subset E_1' \subset \cdots \subset E_{n-1}') \in Y \\ E_0' \in E_{n-1} \setminus E_{n-2};\ E_1' \nparallel E_1,\ E_2' \| [E_1], \cdots, E_{n-1}' \| [E_{n-2}]}}^{(q^n - q^{n-1})} f(E_0' \subset E_1' \subset \cdots \subset E_{n-1}') \quad \text{(cf. } (7_1) \text{ for } f).$$

$$= -(-1)^{n-1} \sum_{\substack{(E_0' \subset E_1' \subset \cdots \subset E_{n-1}') \in Y \\ E_0' \in E_{n-1} \setminus E_{n-2},\ E_1' \| E_1,\ E_2' \| [E_1], \cdots, E_{n-1}' \| [E_{n-2}]}}^{(q^{n-1} - q^{n-2})} f(E_0' \subset E_1' \subset \cdots \subset E_{n-1}')$$

$$= (-1)^n \sum_{\substack{(E_1' \subset \cdots \subset E_{n-1}') \\ E_1' \subset E_{n-1} \setminus E_{n-2},\ E_1' \| E_1 \\ E_2' \| [E_1], \cdots, [E_{n-1}'] \| E_{n-2}}}^{(q^{n-2} - q^{n-3})} \ \sum_{\substack{E_0' \\ E_0' \subset E_1'}}^{(q)} f(E_0' \subset E_1' \subset \cdots \subset E_{n-1}') = 0 \quad \text{(cf. } (7_0) \text{ for } f)$$

and (7_0) for Tf is proved.

We now prove (7_k) for f, $0 < k < n-2$. Let $(E_0 \subset E_1 \subset \cdots \subset E_{k-1} \subset E_{k+1} \subset \cdots \subset E_{n-1})$ be an incomplete flag in V $(0 \notin E_i)$. We have

$$\sum_{\tilde{E}_k}^{(q+1)} (Tf)(E_0 \subset E_1 \subset \cdots \subset E_{k-1} \subset \tilde{E}_k \subset E_{k+1} \subset \cdots \subset E_{n-1})$$

$$= (-1)^{n-1} \sum^{(q^n - q^{n-2})}_{\substack{(E'_0 \subset E'_1 \subset \cdots \subset E'_{n-1}) \in Y \\ E'_0 \in E_{n-1} \setminus E_{n-2},\, E'_1 \| [E_0], \cdots, E'_k \| [E_{k-1}], \\ E'_{k+2} \| [E_{k+1}], \cdots, E'_{n-1} \| [E_{n-2}]}}$$

$$= (-1)^{n-1} \sum^{(q^{n-1} - q^{n-2})}_{\substack{(E'_0 \subset E'_1 \subset \cdots \subset E'_k \subset E'_{k+2} \subset \cdots \subset E'_{n-1}) \\ E'_0 \in E_{n-1} \setminus E_{n-2},\, E'_1 \| [E_0], \cdots, E'_k \| [E_{k-1}], \\ E'_{k+2} \| [E_{k+1}], \cdots, E'_{n-1} \| [E_{n-2}]}} \quad \sum^{(q+1)}_{\substack{E'_{k+1} \\ E'_k \subset E'_{k+1} \subset E'_{k+2}}} f(E'_0 \subset E'_1 \subset \cdots \subset E'_{n-1}) = 0$$

(cf. (7_{k+1}) for f)

and (7_k) for Tf is proved for $0 < k < n-2$.

Next we prove (7_{n-2}) for Tf (here we assume $n \geq 3$). Let $(E_0 \subset E_1 \subset \cdots \subset E_{n-3} \subset E_{n-1})$ be an incomplete flag in V $(0 \notin E_i)$. We have

$$\sum_{\tilde{E}_{n-2}}^{(q+1)} (Tf)(E_0 \subset E_1 \subset \cdots \subset E_{n-3} \subset \tilde{E}_{n-2} \subset E_{n-1})$$

$$= (-1)^{n-1} \sum^{(q^n - q^{n-2})}_{\substack{(E'_0 \subset E'_1 \subset \cdots \subset E'_{n-1}) \in Y \\ E'_0 \in E_{n-1} \setminus E_{n-2},\, E'_1 \| [E_0], \cdots, E'_{n-2} \| [E_{n-3}]}} f(E'_0 \subset E'_1 \subset \cdots \subset E'_{n-1})$$

$$= (-1)^{n-1} \sum^{(q^{n-1} - q^{n-3})}_{\substack{(E'_0 \subset E'_1 \subset \cdots \subset E'_{n-2}) \\ E'_0 \in E_{n-1} \setminus E_{n-2},\, E'_1 \| [E_0], \cdots, E'_{n-2} \| [E_{n-3}]}} \quad \sum^{(q)}_{\substack{E'_{n-1} \\ E'_{n-2} \subset E'_{n-1} \\ 0 \notin E'_{n-1}}} f(E'_0 \subset E'_1 \subset \cdots \subset E'_{n-1}) = 0$$

(cf. (7_{n-1}) for f)

and (7_{n-2}) for Tf is proved.

Finally we prove (7_{n-1}) for Tf. Let $(E_0 \subset E_1 \subset \cdots \subset E_{n-2})$ be an incomplete flag in $V (0 \notin E_i)$. We have

$$\sum_{\substack{\tilde{E}_{n-1} \\ 0 \notin \tilde{E}_{n-1}}}^{(q)} (Tf)(E_0 \subset E_1 \subset \cdots \subset E_{n-2} \subset \tilde{E}_{n-1})$$

$$= (-1)^{n-1} \sum_{\substack{(E'_0 \subset E'_1 \subset \cdots \subset E'_{n-1}) \in Y \\ E'_0 \in V \setminus E_{n-2}, E'_1 \| [E_0], \cdots, E'_{n-1} \| [E_{n-2}]}}^{(q^n - q^{n-1})} f(E'_0 \subset E'_1 \subset \cdots \subset E'_{n-1})$$

$$= (-1)^{n-1} \sum_{\substack{(E'_1 \subset E'_2 \subset \cdots \subset E'_{n-1}) \\ E'_1 \| [E_0], E'_1 \not\subset E_{n-2} \\ E'_2 \| [E_1], \cdots, E'_{n-1} \| [E_{n-2}]}}^{(q^{n-1} - q^{n-2})} \sum_{\substack{E'_0 \\ E'_0 \in E'_1}}^{(q)} f(E'_0 \subset E'_1 \subset \cdots \subset E'_{n-1}) = 0$$

(cf. (7_0) for f)

and the lemma is proved.

It follows from the lemma that $T : \mathcal{F}_A \to \mathcal{F}_A$ takes the subspace $\dot{\Delta}_A(V)$ into itself; the restriction of T to $\dot{\Delta}_A(V)$ will be denoted by the same letter: $T : \dot{\Delta}_A(V) \to \dot{\Delta}_A(V)$.

3.2. In this section we shall prove the following

PROPOSITION. *Let* $f \in \dot{\Delta}_A(V)$ *be such that*

$$f(E_0 \subset E_1 \subset \cdots \subset E_{n-1}) = f(E'_0 \subset E'_1 \subset \cdots \subset E'_{n-1}) \tag{8}$$

whenever $E_0 = E'_0, [E_1] = [E'_1], \cdots, [E_{n-1}] = [E'_{n-1}], (0 \notin E_i, 0 \notin E'_i)$. *Then* $Tf = f$.

Proof. We shall first consider the case $n = 2$. Let $f \in \dot{\Delta}_A(V)$ be satisfying (8). We have

$$(Tf)(E_0 \subset E_1) = - \sum_{\substack{(E_0' \subset E_1') \in Y \\ E_0' \in E_1 \setminus E_0, E_1' \| [E_0]}}^{(q-1)} f(E_0' \subset E_1')$$

$$= - \sum_{E_0' \in E_1 \setminus E_0}^{(q-1)} f(E_0' \subset E_1) \qquad \text{(cf. (8))}$$

$$f(E_0 \subset E_1) \qquad \text{(cf. } (7_0))$$

Let now $n \geq 3$ and let $f \in \dot{\Delta}_A(V)$ be satisfying (8). Let $(E_0 \subset E_1 \subset \cdots \subset E_{n-1}) \in Y$. We shall prove by induction on k the following statement.

(9_k) $(Tf)(E_0 \subset E_1 \subset \cdots \subset E_{n-1})$

$$= (-1)^{n-k-1} \sum_{\substack{E_k' \\ E_{k-1} \subset E_k' \subset E_{n-1} \\ E_k' \not\subset E_{n-2}}} f(E_0 \subset E_1 \subset \cdots \subset E_{k-1} \subset E_k' \subset \langle E_k', E_k \rangle \subset \cdots \subset \langle E_k', E_{n-1} \rangle) \qquad (0 \leq k \leq n-1)$$

Note that (9_{n-1}) is precisely the statement $Tf = f$, and that (9_0) follows from the hypothesis (8). Assume that (9_k) holds for some k, $0 \leq k < n-1$; we shall prove the statement (9_{k+1}). We have

$(Tf)(E_0 \subset E_1 \subset \cdots \subset E_{n-1})$

$$= (-1)^{n-k-1} \sum_{\substack{E_k' \\ E_{k-1} \subset E_k' \subset E_{n-1} \\ E_k' \not\subset E_{n-2}}} f(E_0 \subset E_1 \subset \cdots \subset E_{k-1} \subset E_k' \subset \langle E_k', E_k \rangle \subset \cdots \subset \langle E_k', E_{n-1} \rangle)$$

$$= (-1)^{n-k-1} \sum_{\substack{E'_{k+1} \\ E_k \subset E'_{k+1} \subset E_{n-1} \\ E'_{k+1} \not\subset E_{n-2}}} \sum_{\substack{E'_k \\ E_{k-1} \subset E'_k \subset E'_{k+1} \\ E'_k \neq E_k}} f(E_0 \subset E_1 \subset \cdots \subset E_{k-1} \subset E'_k \subset E'_{k+1} \subset$$

$$< E'_{k+1}, E_{k+1} > \subset \cdots \subset < E'_{k+1}, E_{n-1} >)$$

$$= (-1)^{n-k-2} \sum_{\substack{E'_{k+1} \\ E_k \subset E'_{k+1} \subset E_{n-1} \\ E'_{k+1} \not\subset E_{n-2}}} f(E_0 \subset E_1 \subset \cdots \subset E_k \subset E'_{k+1} \subset < E'_{k+1}, E_{k+1} > \subset \cdots \subset < E'_{k+1}, E_{n-1} >)$$

(cf. (7_k)) and the proposition is proved by induction. (Here we have used the convention $E_{-1} = \emptyset$.)

Note that in case q is invertible in A, any element $f \in \dot{\Delta}_A(V)$ satisfying (8) must be zero.

3.3 In this section we shall assume that $\dim V = 2$. Let $\mathrm{Pol}_{q-2}(V)$ be the set of homogeneous polynomial functions $V \to F$ of degree $(q-2)$. It is clear that $\mathrm{Pol}_{q-2}(V)$ is an F-vector space of dimension $(q-1)$. We shall describe now two F-vector spaces which are canonically isomorphic to $\mathrm{Pol}_{q-2}(V)$ but whose definitions are somewhat more geometric.

First we recall (see 1.13) that $\mathfrak{D}(V)$ is the F-vector space of all functions f which associate to any one-dimensional linear subspace V_1 of V a vector $f(V_1) \in V_1$ such that $\sum_{V_1 \subset V} f(V_1) = 0$ (sum in V). It is clear that $\dim_F \mathfrak{D}(V) = \text{number of lines} - 2 = (q+1) - 2 = q-1$. Let $\mathcal{P}^{(-1)}(V)$ be the F-vector space of all functions $f : V \backslash 0 \to F$ such that $f(\lambda v) = \lambda^{-1} f(v)$, $\lambda \in F^*$ and such that $\sum_{\substack{v \\ v \in E_1}} f(v) = 0$ for any affine hyperplane $E_1 \subset V$, $0 \notin E_1$. Let $f \in \mathrm{Pol}_{q-2}(V)$. f defines a function

$f: V\backslash 0 \to F$ which clearly satisfies $f(\lambda v) = \lambda^{-1} f(v), \lambda \in F^*$. Let E_1 be an affine hyperplane in $V, (0 \notin E_1)$. We can choose a basis e_1, e_2 in V so that $E_1 = \{e_1 + \mu e_2, \mu \in F\}$. We have

$$f(\mu_1 e_1 + \mu_2 e_2) = \sum_{0 \leq i \leq q-2} a_i \mu_1^i \mu_2^{q-2-i}, \quad a_i \in F$$

hence

$$\sum_{\mu \in F} f(e_1 + \mu e_2) = \sum_{0 \leq i \leq q-2} \left(a_i \sum_{\mu \in F} \mu^{q-2-i} \right) = 0$$

(here we have used the identity $\sum_{\mu \in F} \mu^j = 0 \ (0 \leq j \leq q-2)$). This shows that $\sum_{v \in E_1} f(v) = 0$ and hence $f \in \mathcal{P}^{(-1)}(V)$. We have then a natural map $\mathrm{Pol}_{q-2}(V) \to \mathcal{P}^{(-1)}(V)$ which is clearly injective. In order to prove that this is an isomorphism it is then enough to prove that $\mathcal{P}^{(-1)}(V)$ has dimension at most $(q-1)$. To see this, let $E_1 \subset V$, $0 \notin E_1$ and let $v_0 \in E_1$ be fixed. In order to prove that $\dim \mathcal{P}^{(-1)}(V) \leq q-1$ it is sufficient to prove that if $f \in \mathcal{P}^{(-1)}(V)$ is zero on $E_1 \backslash v_0$ then f must be identically zero. Consider such a function f. We must clearly have $f(v_0) = 0$ since $\sum_{v \in E_1} f(v) = 0$. From the homogeneity property of f it follows that $f(v) = 0$ for any $v \in V \backslash V_1$ where V_1 is the line through 0, parallel to E_1. Let $v_1 \in V_1 \backslash 0$. Let E_1' be an affine hyperplane in V, such that $v_1 \in E_1'$, $0 \notin E_1'$. Then $f(v) = 0$ for all points v of E_1', other than v_1. It follows that $f(v_1) = 0$, and hence the assertion that $\mathrm{Pol}_{q-2}(V) \to \mathcal{P}^{(-1)}(V)$ is an isomorphism is proved.

Let now $f \in \mathfrak{D}(V)$. We associate to f a function $\tilde{f}: V-0 \to F$ defined by the formula $\tilde{f}(v) v = f[v]$ (here $[v]$ is the linear span of v). It is clear that $\tilde{f}(\lambda v) = \lambda^{-1} \tilde{f}(v), \lambda \in F^*$. Let $E_1 \subset V$, $0 \notin E_1$ be an affine hyperplane. We shall show that $\sum_{v \in E_1} \tilde{f}(v) = 0$. In fact, let v^* be the unique linear function $V \to F$ such that $v^*(v) = 1$ for $v \in E_1$. We have $\sum_{v \in E_1} f[v] + f(\ker v^*) = 0$, since $f \in \mathfrak{D}(V)$. Applying v^* to this equality we get $\sum_{v \in E_1} v^*(f[v]) = 0$ and hence $\sum_{v \in E_1} \tilde{f}(v) v^*(v) = \sum_{v \in E_1} \tilde{f}(v) = 1$. It follows that the correspondence $f \to \tilde{f}$ defines a map $\mathfrak{D}(V) \to \mathcal{P}^{(-1)}(V)$. This is clearly injective, and since the two vector spaces involved have the same dimension, it must be an isomorphism. We collect these results in the following

PROPOSITION. *If* V *is a 2-dimensional vector space over* F, *there exist canonical isomorphisms of* F*-vector spaces*

$$\mathrm{Pol}_{q-2}(V) \approx \mathcal{P}^{(-1)}(V) \approx \mathfrak{D}(V) .$$

3.4 In this section we shall assume that $A = F$. Define $\dot{\Delta}_F(V)^{(k)}$ as the set of all $f \in \dot{\Delta}_F(V)$ such that $f(\lambda E_0 \subset \lambda E_1 \subset \cdots \subset \lambda E_{n-1}) = \lambda^k f(E_0 \subset E_1 \subset \cdots \subset E_{n-1})$ for all $\lambda \in F^*$ and all $(E_0 \subset E_1 \subset \cdots \subset E_{n-1}) \in Y$. Here k is some fixed element in $\mathbf{Z}/(q-1)\mathbf{Z}$. It is obvious that

$$\dot{\Delta}_F(V) = \bigoplus_{k \in \mathbf{Z}/(q-1)\mathbf{Z}} \dot{\Delta}_F(V)^{(k)} .$$

Note that the endomorphism T of $\dot{\Delta}_F(V)$ leaves invariant each of the subspaces $\dot{\Delta}_F(V)^{(k)}$.

We now prove the following partial converse to Proposition 3.2.

PROPOSITION. *Let* $f \in \dot{\Delta}_F(V)^{(-1)}$ *be such that* $Tf = f$. *Then* f *satisfies condition* (8) *of* 3.2.

Proof. Let $\varepsilon = (E_0 \subset E_1 \subset \cdots \subset E_{n-1})$, $\tilde{\varepsilon} = (\tilde{E}_0 \subset \tilde{E}_1 \subset \cdots \subset \tilde{E}_{n-1})$ be two elements of Y such that $E_0 = \tilde{E}_0, [E_1] = [\tilde{E}_1], \cdots, [E_{n-1}] = [\tilde{E}_{n-1}]$. We wish to prove that $f(\varepsilon) = f(\tilde{\varepsilon})$. Since $Tf = f$, this is equivalent to

$$\sum_{\substack{(E'_0 \subset \cdots \subset E'_{n-1}) \epsilon Y \\ E'_0 \epsilon E_{n-1} \setminus E_{n-2} \\ E'_1 \| [E_0], \cdots, E'_{n-1} \| [E_{n-2}]}} f(E'_0 \subset E'_1 \subset \cdots \subset E'_{n-1}) = \sum_{\substack{(E''_0 \subset E''_1 \subset \cdots \subset E''_{n-1}) \epsilon Y \\ E''_0 \epsilon \tilde{E}_{n-1} \setminus \tilde{E}_{n-2} \\ E''_1 \| [\tilde{E}_0], \cdots, E''_{n-1} \| [\tilde{E}_{n-2}]}} f(E''_0 \subset E''_1 \subset \cdots \subset E''_{n-1}).$$

We define a function $\phi : V \setminus [E_{n-2}] \to F$ by $\phi(E'_0) = f(E'_0 \subset E'_1 \subset \cdots \subset E'_{n-1})$ where $E'_i (1 \leq i \leq n-1)$ is the unique i-dimensional affine subspace of V such that $E'_0 \epsilon E'_i$ and $E'_i \| [E_{i-1}]$. Then the equality to be proved can be written as

$$\sum_{E'_0 \epsilon E'_{n-1} \setminus E'_{n-1} \cap [E_{n-2}]} \phi(E'_0) = \sum_{E''_0 \epsilon E''_{n-1} \setminus E''_{n-1} \cap [E_{n-2}]} \phi(E''_0)$$

This clearly follows from the following

LEMMA. *Let* V *be a vector space of dimension* $n \geq 2$ *over* F, V_{n-1} *a linear hyperplane in* V *and* E'_{n-1}, E''_{n-1} *two affine hyperplanes in* V *not containing* 0. *Let* $E_0 \epsilon E'_{n-1} \cap E''_{n-1} \cap V_{n-1}$. *Assume that* $\phi : V \setminus V_{n-1} \to F$ *is a function such that* $\phi(\lambda E'_0) = \lambda^{-1} \phi(E'_0)$, $\lambda \epsilon F^*$, $E'_0 \epsilon V \setminus V_{n-1}$ *and*

$$\sum_{\substack{E'_0 \\ E'_0 \subset E'_1}}^{(q)} \phi(E'_0) = 0$$

for any one-dimensional affine subspace $E'_1 \subset V$ $(E'_1 \| [E_0], E'_1 \not\subset V_{n-1})$. *Then*

$$\sum_{E'_0 \epsilon E'_{n-1} \setminus (E'_{n-1} \cap V_{n-1})}^{(q^{n-1} - q^{n-2})} \phi(E'_0) = \sum_{E''_0 \epsilon E''_{n-1} \setminus (E''_{n-1} \cap V_{n-1})}^{(q^{n-1} - q^{n-2})} \phi(E''_0) .$$

Proof of the Lemma. Let $\mathcal{X}$ be the set of all 2-dimensional linear subspaces π of V such that $\pi \supset [E_0]$, $\pi \not\subset V_{n-1}$. The equality to be proved can be written as

$$\sum_{\pi\in\mathcal{X}}^{(q^{n-2})} \sum_{E'_0\in(\pi\cap E'_{n-1})\setminus E_0}^{(q-1)} \phi(E'_0) = \sum_{\pi\in\mathcal{X}}^{(q^{n-2})} \sum_{E''_0\subset(\pi\cap E''_{n-1})\setminus E_0}^{(q-1)} \phi(E''_0)$$

so that the general case would follow from the 2-dimensional case. Assume now that $\dim V = 2$. We wish to prove that

$$\sum_{E'_0\in E'_1\setminus E_0}^{(q-1)} \phi(E'_0) = \sum_{E''_0\in E''_1\setminus E_0}^{(q-1)} \phi(E''_0) .$$

Let $\mathcal{S}$ be the set of all functions $\phi : V - [E_0] \to F$ satisfying the hypothesis of the Lemma $(E_0, E'_1, E''_1, V_1 = [E_0]$ fixed). It is obvious that $\mathcal{S}$ is an F-vector space of dimension $(q-1)$. On the other hand, there is a natural restriction map $\mathcal{P}^{(-1)}(V) \to \mathcal{S}$ (see 3.3). This map is clearly injective (a function $f \in \mathcal{P}^{(-1)}(V)$ which vanishes on $V - [E_0]$ must be identically zero, cf. 3.3). Since the vector spaces involved have dimension $q-1$ the map $\mathcal{P}^{(-1)}(V) \to \mathcal{S}$ must be isomorphism.

Our function ϕ belongs to $\mathcal{S}$ by hypothesis. From what we have just proved, it follows that there exists $f \in \mathcal{P}^{(-1)}(V)$ such that $f(E'_0) = \phi(E'_0)$ for all $E'_0 \in V\setminus[E_0]$. It follows that

$$\sum_{E'_0\in E'_1\setminus E_0}^{(q-1)} \phi(E'_0) = f(E_0)$$

and similarly

$$\sum_{E''_0\in E''_1\setminus E_0}^{(q-1)} \phi(E''_0) = f(E_0)$$

and the Lemma is proved.

3.5. From 3.2 and 3.4 we deduce that $T:\dot{\Delta}_F(V)^{(-1)} \to \dot{\Delta}_F(V)^{(-1)}$ is idempotent and its image is precisely

$$\mathcal{T} = \{f \in \dot{\Delta}_F(V)^{(-1)} \mid f \text{ satisfies (8)}\} .$$

In this section we shall construct a canonical isomorphism $\mathcal{T} \cong \mathfrak{D}(V)$. We recall (cf. 1.13) that $\mathfrak{D}(V)$ is the set of all functions ϕ which associate to any complete linear flag $(V_1 \subset V_2 \subset \cdots \subset V_{n-1})$ in V a vector $\phi(V_1 \subset V_2 \subset \cdots \subset V_{n-1}) \in V_1$ such that the following conditions are satisfied

$$(10_i) \qquad \sum_{\tilde{V}_i}^{(q+1)} \phi(V_1 \subset V_2 \subset \cdots \subset V_{i-1} \subset \tilde{V}_i \subset V_{i+1} \subset \cdots \subset V_{n-1}) = 0$$

$(1 \leq i \leq n-1)$ where $(V_1 \subset V_2 \subset \cdots \subset V_{i-1} \subset V_{i+1} \subset \cdots \subset V_{n-1})$ is any given incomplete flag (the empty flag if $n=2$, $i=1$). Note that this is a sum of vectors in V_1 if $i > 1$ or in V_2 if $i = 1$.

We define a map $\alpha : \mathcal{T} \to \mathfrak{D}(V)$, as follows. Given $f \in \mathcal{T}$, define $\alpha f \in \mathfrak{D}(V)$ by the formula

$$(\alpha f)(V_1 \subset V_2 \subset \cdots \subset V_{n-1}) = f(E_0 \subset E_1 \subset \cdots \subset E_{n-1}) E_0$$

where $(E_0 \subset E_1 \subset \cdots \subset E_{n-1})$ is any element of Y such that $E_0 \in V_1 \setminus 0$, $E_1 \subset V_2 \setminus 0, \cdots, E_{n-1} \subset V \setminus 0$. (This is independent of the choice since f satisfies (8) and is homogeneous of degree (-1).) We must check that αf satisfies the conditions (10_i) $(1 \leq i \leq n-1)$. If $i \geq 2$ this follows easily from the condition (7_{i-1}) satisfied by f. In order to check (10_1) for αf we can clearly assume that $\dim V = 2$. In this case we must prove that if $f \in \mathcal{P}^{(-1)}(V)$ (see 3.3) then the function αf defined by $\alpha f(V_1) = f(E_0) E_0$, $E_0 \in V_1 \setminus 0$ lies in $\mathfrak{D}(V)$; but this follows from the isomorphism $\mathfrak{D}(V) \overset{\approx}{\to} \mathcal{P}^{(-1)}(V)$ constructed in 3.3.

Next, we define a map $\beta : \mathfrak{D}(V) \to \mathcal{T}$. Given $\phi \in \mathfrak{D}(V)$ define $\beta\phi \in \mathcal{T}$ by the formula:

$$(\beta\phi)(E_0 \subset E_1 \subset \cdots \subset E_{n-1})E_0 = \phi([E_0] \subset [E_1] \subset \cdots \subset [E_{n-2}]),$$
$$(E_0 \subset E_1 \subset \cdots \subset E_{n-1}) \in Y.$$

It is clear that $\beta\phi$ is homogeneous of degree (-1) and that $\beta\phi$ satisfies (8). We must check now that $\beta\phi$ satisfies the conditions (7_i), $(0 \leq i \leq n-1)$. If $0 < i < n-1$ this follows easily from the condition (10_{i+1}) satisfied by ϕ. (7_{n-1}) is just $q\phi([E_0] \subset [E_1] \subset \cdots \subset [E_{n-2}]) = 0$. In order to prove (7_0) we can again assume that $\dim V = 2$ and the result follows from the isomorphism $\mathcal{D}(V) \approx \mathcal{P}^{(-1)}(V)$ constructed in 3.3.

It is obvious that $a \circ \beta = 1_{\mathcal{D}(V)}$ and $\beta \circ a = 1_{\mathcal{T}}$. We deduce the following

THEOREM. *Assume* $\dim_F V \geq 2$. *Then there is canonical isomorphism of* F*-vector spaces*

$$\beta : \mathcal{D}(V) \approx \{f \in \dot{\Delta}_F(V)^{(-1)} \mid Tf = f\}\ .$$

β *is uniquely characterized by the property*

$$(\beta\phi)(E_0 \subset E_1 \subset \cdots \subset E_{n-1})E_0 = \phi([E_0] \subset [E_1] \subset \cdots \subset [E_{n-2}])$$

for all $\phi \in \mathcal{D}(V), (E_0 \subset E_1 \subset \cdots \subset E_{n-1}) \in Y$.

3.6 We recall now the definition of the ring of Witt vectors W_F associated to the finite field F. Let p be the characteristic of F. W_F is a commutative ring together with a ring homomorphism $W_F \to F$, which is uniquely characterized by the following properties (see [10], p. 48).

a. p is not a zero divisor in W_F
b. W_F is Hausdorff and complete with respect to the topology determined by the ideals $p^m W_F (m \geq 1)$.
c. The sequence $0 \to pW_F \to W_F \to F \to 0$ is exact.

It is well known that W_F must be in fact an integral domain of characteristic zero, that there exists a unique map $F^* \to W_F$ (denoted

$\lambda \to \tilde{\lambda}$) such that $\tilde{\lambda} \cdot \tilde{\mu} = \widetilde{\lambda\mu}(\lambda, \mu \in F^*)$ and the composition $F^* \to W_F \to F$ is the natural inclusion $F^* \to F$. ($\lambda \to \tilde{\lambda}$ is known as the Teichmüller map.)

W_F can be described explicitly as the set of all sequences $(\lambda_0, \lambda_1, \lambda_2, \cdots)$ of elements in F in which the addition and multiplication are given by certain universal polynomials discovered by Witt (see [10], p. 49).

If F is the prime field of characteristic p, W_F is canonically isomorphic to the ring of p-adic integers.

If $F \to F'$ is an imbedding of a finite field in another, there is a corresponding imbedding $W_F \to W_{F'}$. Moreover the Galois group $\mathrm{Gal}(F'/F)$ acts naturally on $W_{F'}$ so that W_F is precisely the ring of invariants.

3.7 In this section we shall prove the following

LEMMA. *Let* M *be a finitely generated free* W_F*-module* (F *a finite field*). *Let* $T: M \to M$ *be a* W_F*-linear map such that* $T \underset{W_F}{\otimes} 1_F : M \underset{W_F}{\otimes} F \to M \underset{W_F}{\otimes} F$ *is idempotent. Let*

$$M' = \{x \in M \mid \lim_{i\to\infty} (1-T)^i x = 0\},\ M'' = \{x \in M \mid \lim_{i\to\infty} T^i x = 0\} .$$

Then $M = M' \oplus M''$.

Proof. We shall prove the lemma assuming first the following statement: there exists $T' \in \mathrm{End}_{W_F}(M)$ such that $T'^2 = T'$, $T'T = TT'$ and $T \underset{W_F}{\otimes} 1_F = T' \underset{W_F}{\otimes} 1_F$. Then we have clearly $M = \ker(1-T') \oplus \ker T'$ and T commutes with this decomposition. Moreover there exist $T_1 \in \mathrm{End}_{W_F}(\ker(1-T'))$ and $T_2 \in \mathrm{End}_{W_F}(\ker T')$ such that $T = 1 - pT_1$ on $\ker(1-T')$ and $T = pT_2$ on $\ker T'$. Let $x \in M$; we can write uniquely $x = x_1 + x_2, x_1 \in \ker(1-T')$, $x_2 \in \ker T'$. Then

$$(1-T)^i x = (pT_1)^i x_1 + (1-pT_2)^i x_2 .$$

Since $\lim_{i\to\infty}(1-T)^i x = 0$ and $\lim_{i\to\infty}(pT_1)^i x_1 = 0$ we must have $\lim_{i\to\infty}(1-pT_2)^i x_2 = 0$, hence $x_2 = 0$. This shows that $M' \subset \ker(1-T')$. The reverse inclusion is obvious hence $M' = \ker(1-T')$. Similarly one proves that $M'' = \ker T'$ and it follows that we have the direct sum decomposition $M = M' \oplus M''$. The existence of T' is a classical fact about "lifting idempotents". The idea is to construct T' by a limit procedure. Define $Y_0 = T$, $X_0 = Y_0^2 - Y_0, \cdots, Y_{i+1} = Y_i + X_i - 2Y_i X_i$, $X_{i+1} = Y_{i+1}^2 - Y_{i+1}$ $(i \geq 0), \cdots$. One checks by induction that X_i is divisible by p^{2^i}. It follows that $Y_{i+1} - Y_i = X_i(1-2Y_i)$ is divisible by p^{2^i} hence Y_i converges to some $T' \in \mathrm{End}_{W_F}(M)$ and it is easy to see that $T'^2 = T'$, $TT' = T'T$ and $T \otimes_{W_F} 1_F = T' \otimes_{W_F} 1_F$. The first part of the proof shows that T' is in fact uniquely determined by these properties.

3.8 We use the notations of 3.5. We shall apply the Corollary 3.7 in the following situation. We define $\dot{\Delta}_{W_F}(V)^{(k)}$ as the set of all $f \in \dot{\Delta}_{W_F}(V)$ satisfying the "homogeneity condition"

$$f(\lambda E_0 \subset \lambda E_1 \subset \cdots \subset \lambda E_{n-1}) = \tilde{\lambda}^k f(E_0 \subset E_1 \subset \cdots \subset E_{n-1})$$

for all $\lambda \in F^*$ and all $(E_0 \subset E_1 \subset \cdots \subset E_{n-1}) \in \dot{Y}$. Here k is some fixed element in $\mathbf{Z}/(q-1)\mathbf{Z}$. We have $\dot{\Delta}_{W_F}(V) = \bigoplus_{k \in \mathbf{Z}/(q-1)\mathbf{Z}} \dot{\Delta}_{W_F}(V)^{(k)}$ and $\dot{\Delta}_F(V)^{(k)} = \dot{\Delta}_{W_F}(V)^{(k)} \otimes_{W_F} F$.

Let $T : \dot{\Delta}_{W_F}(V)^{(-1)} \to \dot{\Delta}_{W_F}(V)^{(-1)}$ be the restriction of the map $T : \dot{\Delta}_{W_F}(V) \to \dot{\Delta}_{W_F}(V)$, defined in 3.1 (note that all the subspaces $\dot{\Delta}_{W_F}(V)^{(k)}$ are invariant under T).

It is obvious that we have a commutative diagram

$$\begin{array}{ccc} \dot{\Delta}_{W_F}(V)^{(-1)} & \xrightarrow{T} & \dot{\Delta}_{W_F}(V)^{(-1)} \\ \downarrow & & \downarrow \\ \dot{\Delta}_F(V)^{(-1)} & \xrightarrow{T} & \dot{\Delta}_F(V)^{(-1)} \end{array}$$

where the vertical arrows are the natural projections induced by $W_F \to F$. Since T in the bottom row is idempotent (cf. 3.5), we can apply Lemma 3.7 and deduce that we have a direct sum decomposition

$$\dot{\Delta}_{W_F}(V)^{(-1)} = D(V) \oplus D(V)^{\perp}$$

where, by definition,

$$D(V) = \{f \in \dot{\Delta}_{W_F}(V)^{(-1)} \mid \lim_{i\to\infty} (1-T)^i f = 0\}$$

and

$$D(V)^{\perp} = \{f \in \dot{\Delta}_{W_F}(V)^{(-1)} \mid \lim_{i\to\infty} T^i f = 0\} .$$

By tensoring with F we get a direct sum decomposition

$$\dot{\Delta}_F(V)^{(-1)} = (D(V) \underset{W_F}{\otimes} F) \oplus (D(V) \underset{W_F}{\otimes} F)$$

where

$$D(V) \underset{W_F}{\otimes} F = \{f \in \dot{\Delta}_F(V)^{(-1)} \mid (1-T)f = 0\}$$

and

$$D(V)^{\perp} \underset{W_F}{\otimes} F = \{f \in \dot{\Delta}_F(V)^{(-1)} \mid Tf = 0\} .$$

Note that $D(V) \underset{W_F}{\otimes} F$ is precisely the subspace of $\dot{\Delta}_F(V)^{(-1)}$ which in 3.5 was denoted by $\mathcal{T}$. It follows from 3.5 that there is a canonical isomorphism of F-vector spaces $\beta : \mathfrak{D}(V) \approx D(V) \underset{W_F}{\otimes} F$ defined by

$$(\beta\phi)(E_0 \subset E_1 \subset \cdots \subset E_{n-1})E_0 = \phi([E_0] \subset [E_1] \subset \cdots \subset [E_{n-2}])$$

for all $(E_0 \subset E_1 \subset \cdots \subset E_{n-1}) \in Y$, $\phi \in \mathfrak{D}(V)$.

We shall call D(V) *the distinguished discrete series module associated to* V. Note that we have defined D(V) only in case $\dim_F V \geq 2$. We can extend the definition of D(V) to the case when $\dim_F(V) = 1$. In this case we define

$$D(V) = \{f : V \setminus 0 \to W_F \mid f(\lambda E_0) = \tilde{\lambda}^{-1} f(E_0), E_0 \in V \setminus 0, \lambda \in F^*\}.$$

This is a free W_F-module of rank 1 and there is an obvious isomorphism $\beta : \mathfrak{D}(V) \to D(V) \underset{W_F}{\otimes} F$ defined by $\beta(v)(E'_0) = \frac{v}{E'_0}$ where $v \in \mathfrak{D}(V) = V$ and $E'_0 \in V \setminus 0$. If $n \geq 2$, $D(V)$ is a free W_F-module of rank $(q-1)(q^2-1)\cdots(q^{n-1}-1)$. This follows from the corresponding formula for $\dim_F \mathfrak{D}(V)$, see 1.14.

3.8 Let E be an affine hyperplane in V such that $0 \notin E$, $(\dim V \geq 2)$. In this section we shall prove that E gives rise to a canonical W_F-isomorphism $\rho_E : D(V) \approx \Delta_{W_F}(E)$. Define ρ_E by the formula

$$(\rho_E f)(E_0 \subset E_1 \subset \cdots \subset E_{n-2}) = f(E_0 \subset E_1 \subset \cdots \subset E_{n-2} \subset E)$$

for any complete affine flag $(E_0 \subset E_1 \subset \cdots \subset E_{n-2})$ in E and any $f \in D(V)$. It is clear that ρ_E is an isomorphism if and only if $\rho_E \underset{W_F}{\otimes} 1_F$ is an isomorphism. Since $\dim_F(D(V) \underset{W_F}{\otimes} F) = \dim_F(\Delta_F(E)) = (q-1)(q^2-1)\cdots(q^{n-1}-1)$ it is sufficient to prove that $\rho_E \underset{W_F}{\otimes} 1_F$ is injective. By composing with the isomorphism $D(V)_{W_F} \otimes F \cong \mathfrak{D}(V)$, $\rho_E \underset{W_F}{\otimes} 1_F$ becomes the map $\bar{\rho}_E : \mathfrak{D}(V) \to \Delta_F(V)$ given by

$$(\bar{\rho}_E \phi)(E_0 \subset E_1 \subset \cdots \subset E_{n-2}) E_0 = \phi([E_0] \subset [E_1] \subset \cdots \subset [E_{n-2}])$$

$\phi \in \mathfrak{D}(V)$, (see 3.7).

Let $\phi \in \mathfrak{D}(V)$ be such that $\bar{\rho}_E \phi = 0$. We shall prove by induction on i that $\phi(V_1 \subset V_2 \subset \cdots \subset V_{n-1}) = 0$ whenever $V_i \cap E \neq \emptyset$. ($i$ fixed, $1 \leq i \leq n$; use the convention $V_n = V$.) For $i = 1$, this is just the assumption $\bar{\rho}_E \phi = 0$. Let $(V_1 \subset V_2 \subset \cdots \subset V_{n-1})$ be a complete linear flag in V such that $V_k \cap E \neq \emptyset$, $V_{k-1} \cap E = \emptyset$ for some k, $1 < k < n$. We have then by the cycle condition for ϕ:

$$\phi(V_1 \subset V_2 \subset \cdots \subset V_{k-2} \subset V_{k-1} \subset V_k \subset \cdots \subset V_{n-1})$$

$$= - \sum_{\tilde{V}_{k-1}}^{(q)} \phi(V_1 \subset V_2 \subset \cdots \subset V_{k-2} \subset \tilde{V}_{k-1} \subset V_k \subset \cdots \subset V_{n-1})$$

where the sum is over all $\tilde{V}_{k-1}$ such that $V_{k-2} \subset \tilde{V}_{k-1} \subset V_k$ and $\tilde{V}_{k-1} \neq V_k$. It is clear that all these subspaces $\tilde{V}_{k-1}$ satisfy $\tilde{V}_{k-1} \cap E \neq \emptyset$ hence the induction hypothesis can be applied. It follows that $\phi(V_1 \subset V_2 \subset \cdots \subset V_{n-1}) = 0$ whenever $V_n \cap E \neq \emptyset$ which is always satisfied. Hence $\phi = 0$. This proves that $\bar{\rho}_E$ is injective hence $\rho_E \otimes 1_F$ is injective and we can state the following

THEOREM. *Assume* $\dim V \geq 2$, *and let* E *be an affine hyperplane in* V *such that* $0 \notin E$. *Then there exists a canonical isomorphism of* W_F*-modules* $\rho_E : D(V) \approx \Delta_{W_F}(E)$ *given by the formula*

$$(\rho_E f)(E_0 \subset E_1 \subset \cdots \subset E_{n-2}) = f(E_0 \subset E_1 \subset \cdots \subset E_{n-2} \subset E)$$

for any complete flag $(E_0 \subset E_1 \subset \cdots \subset E_{n-2})$ *in* E *and* $f \in D(V)$.

3.9 Let $GL(V)$ be the group of all F-vector space isomorphisms $t : V \approx V$. $GL(V)$ acts on $\dot{\Delta}_A(V)$ by the formula

$$(tf)(E_0 \subset E_1 \subset \cdots \subset E_{n-1}) = f(t^{-1}E_0 \subset t^{-1}E_1 \subset \cdots \subset t^{-1}E_{n-1}) .$$

It is clear that the subspace $D(V)$ of $\dot{\Delta}_{W_F}(V)$ is invariant under $GL(V)$ because its definition is intrinsic. Similarly, $\mathfrak{D}(V)$ is a $GL(V)$-module.

Now let $W_F \to \Omega$ be a ring homomorphism of W_F into some algebraically closed field Ω (of characteristic zero). According to 3.8 we have then a canonical isomorphism of Ω-vector spaces $\rho_E \underset{W_F}{\otimes} 1_\Omega : D(V) \underset{W_F}{\otimes} \Omega \approx \Delta_\Omega(E)$. But $\Delta_\Omega(E)$ is an irreducible $Aff(E)$-module (cf. 2.1). Hence the GL(V)-module $D(V) \otimes_{W_F} \Omega$ is irreducible when restricted to the subgroup Aff(E) of GL(V). It follows that $D(V) \otimes_{W_F} \Omega$ *must be irreducible as a GL(V)-module*. Consider now the map $T \in \mathrm{End}_{W_F}(\dot{\Delta}_{W_F}(V)^{(-1)})$ of 3.8. We must clearly have $T(D(V)) \subset D(V)$, hence T defines a GL(V)-endomorphism of D(V). From the irreducibility of the GL(V)-module $D(V) \otimes_{W_F} \Omega$ and the Schur lemma, it follows that there exists a unique Witt vector $\lambda(V) \in W_F$ such that $Tf = \lambda(V)f$, for all $f \in D(V)$. Moreover since $T \otimes_{W_F} 1_F$ is the identity on $D(V) \otimes_{W_F} F$, we must have $\lambda(V) \equiv 1 (\mathrm{mod}\ p)$.

Let $f \in \dot{\Delta}_{W_F}(V)^{(-1)}$ be such that $f \neq 0$ and $Tf = \mu f$ for some $\mu \in W_F$, $\mu \equiv 1\ (\mathrm{mod}\ p)$. Then clearly $\lim_{i \to \infty} (1-T)^i f = 0$ hence $f \in D(V)$ and $\mu = \lambda(V)$.

We have proved the following

PROPOSITION. *Consider the W_F-endomorphism T of $\dot{\Delta}_{W_F}(V)^{(-1)}$. Then there exists a unique Witt vector $\lambda = \lambda(V) \in W_F$ such that $\lambda \equiv 1$ (mod p) and such that there exists $f \in \dot{\Delta}_{W_F}(V)^{(-1)}$, $f \neq 0$ with $Tf = \lambda f$. Moreover, we have*

$$D(V) = \{f \in \dot{\Delta}_{W_F}(V)^{(-1)} \mid Tf = \lambda(V) f\} .$$

Note that this can be taken as a definition of D(V). The Witt vector $\lambda(V)$ will be determined explicitly in 4.8. The W_F-module $D(V)^{\perp}$ can be also described in terms of $\lambda(V)$. We have

$$D(V)^{\perp} = \mathrm{Image}\,(T - \lambda(V) : \dot{\Delta}_{W_F}(V)^{(-1)} \to \dot{\Delta}_{W_F}(V)^{(-1)}) .$$

In fact, let $f = f_1 + f_2, f_1 \in D(V), f_2 \in D(V)^{\perp}$. Then $(T-\lambda(V))f = (T-\lambda(V))f_2 \in D(V)^{\perp}$. Hence it is enough to prove that $T-\lambda(V) : D(V)^{\perp} \to D(V)^{\perp}$ is an isomorphism. This follows by observing that

$$(T - \lambda(V)) \otimes_{W_F} 1_F = -1 \text{ on } D(V)^{\perp} \otimes_{W_F} 1_F .$$

3.10 Let V be a vector space of dimension $n \geq 3$ over F and let $\tilde{V}$ be an m-dimensional linear subspace of V $(n-1 \geq m \geq 1)$. In this section we shall construct a canonical W_F-isomorphism $\psi : D(V/\tilde{V}) \underset{W_F}{\otimes} P_{W_F}(V,\tilde{V}) \approx D(V)$. Let $\mathcal{H}(V,\tilde{V})$ be the set of all affine hyperplanes E in V such that $0 \notin E$ and $\tilde{V}$ is parallel to E. Assume first that $n-2 \geq m \geq 1$. Let $E \in \mathcal{H}(V,\tilde{V})$. Consider the diagram

$$\begin{array}{ccc}
D(V/\tilde{V}) \underset{W_F}{\otimes} P_{W_F}(V,\tilde{V}) & \xrightarrow[\approx]{\tilde{\psi}_E} & D(V) \\
\rho_{E/\tilde{V}} \otimes 1 \Big\downarrow \wr & & \wr \Big\downarrow \rho_E \\
\Delta_{W_F}(E/\tilde{V}) \underset{W_F}{\otimes} P_{W_F}(V,\tilde{V}) & \xrightarrow[\approx]{\psi_E} & \Delta_{W_F}(E)
\end{array}$$

where ρ_E, $\rho_{E/\tilde{V}}$ are the isomorphisms constructed in 3.8, ψ_E is the isomorphism constructed in 2.4 (where also $E/\tilde{V}$ is defined); $\tilde{\psi}_E$ is defined so that this diagram commutes.

We now study the map $\tilde{\psi}_E \underset{W_F}{\otimes} 1_F = \bar{\psi}_E$. It follows easily from the definitions that

$$(11) \qquad \bar{\psi}_E(\phi \otimes u)(V_1 \subset V_2 \subset \cdots \subset V_{n-1}) = \phi(\pi V_1 \subset \pi V_2 \subset \cdots \subset \pi V_{n-m-1}) \cdot u(V_{n-m} \subset V_{n-m+1} \subset \cdots \subset V_{n-1})$$

where $(V_1 \subset V_2 \subset \cdots \subset V_{n-1})$ is any complete linear flag in V such that $V_i \cap E \neq \emptyset$ $(1 \leq i \leq n-1)$ and $V_i + \tilde{V} = V$ $(n-m \leq i \leq n-1)$; we have $\phi \in \mathfrak{D}(V/\tilde{V})$, $u \in P_F(V,\tilde{V})$, $\tilde{\psi}_E(\phi \otimes u) \in \mathfrak{D}(V)$, and π is the natural projection $V \to V/\tilde{V}$.

It is easy to see that (11) remains valid when we only assume that $V_i + \tilde{V} = V$ $(n-m \leq i \leq n-1)$ (we drop the assumption $V_i \cap E \neq \emptyset$). This is proved by induction in exactly the same way as in the induction proof in 3.8, using the cycle conditions for $\bar{\psi}_E(\phi \otimes u) \in \mathfrak{D}(V)$.

We now show that given $E, E' \in \mathcal{H}(V, \tilde{V})$ we have $\overline{\psi}_E = \overline{\psi}_{E'}$. It is clear from the above refinement of (11) that $(\overline{\psi}_E - \overline{\psi}_{E'})(\phi \otimes u)(V_1 \subset V_2 \subset \cdots \subset V_{n-1}) = 0$ provided $V_i + \tilde{V} = V$ $(n-m \leq i \leq n-1)$.

Now, Lemma 2.4 implies obviously that if $\phi_1 \in \mathfrak{D}(V)$ is such that $\phi_1(V_1 \subset V_2 \subset \cdots \subset V_{n-1}) = 0$ whenever $V_i + \tilde{V} = V$ $(n-m \leq i \leq n-1)$ then $\phi_1 = 0$. It follows that $(\overline{\psi}_E - \overline{\psi}_{E'})(\phi \otimes u) = 0$ hence $\overline{\psi}_E = \overline{\psi}_{E'}$ for $E, E' \in \mathcal{H}(V, \tilde{V})$.

We now define a W_F-linear map $\psi : D(V/\tilde{V}) \underset{W_F}{\otimes} P_{W_F}(V, \tilde{V}) \to D(V)$ by the formula

$$\psi = - \sum_{E \in \mathcal{H}(V, \tilde{V})} \tilde{\psi}_E \,.$$

This map is determined canonically by V and $\tilde{V}$. To prove that ψ is an isomorphism, it is enough to prove that $\psi \underset{W_F}{\otimes} 1_F$ is an isomorphism. But under reduction mod p, all the maps $\tilde{\psi}_E$ become equal hence we have

$$\psi \underset{W_F}{\otimes} 1_F = - \sum_{E \in \mathcal{H}(V . \tilde{V})} \overline{\psi}_E = -(q^{n-m}-1)\overline{\psi}_{E_0} = \overline{\psi}_{E_0}$$

where E_0 is a fixed element of $\mathcal{H}(V, \tilde{V})$. (Note that the set $\mathcal{H}(V, \tilde{V})$ has cardinal $(q^{n-m}-1)$.) Since $\overline{\psi}_{E_0}$ is an isomorphism, it follows that $\psi \underset{W_F}{\otimes} 1_F$ and ψ are isomorphisms.

Finally we assume that $\dim V = n \geq 2$ and $\dim \tilde{V} = n-1$. We shall prove that in this case we still have a canonical W_F-isomorphism $\psi : D(V/\tilde{V}) \underset{W_F}{\otimes} P_{W_F}(V, \tilde{V}) \approx D(V)$. The proof is similar to the above.

Given $E \in \mathcal{H}(V, \tilde{V})$ we have a canonical basis $\pi(E) \in V/\tilde{V}$ hence we get a natural isomorphism $\rho_{E/\tilde{V}} : D(V/\tilde{V}) \approx W_F$. Define $\tilde{\psi}_E$ so that the diagram

$$\begin{array}{ccc} D(V/\tilde V)\underset{W_F}{\otimes} P_{W_F}(V,\tilde V) & \xrightarrow[\approx]{\tilde\psi_E} & D(V)\\ \rho_{E/\tilde V}\otimes 1 \Big\downarrow \wr & & \wr \Big\downarrow \rho_E \\ W_F \underset{W_F}{\otimes} P_{W_F}(V,\tilde V) & \xrightarrow[\approx]{\psi_E} & \Delta_{W_F}(E)\end{array}$$

commutes, where ψ_E is the obvious isomorphism, coming from the isomorphism of partially ordered sets $S_{IV}(V,\tilde V) \approx S_I(E)$ (see the remark in 1.11). It is easy to see that $\tilde\psi_E$ is independent of E (not only its reduction mod p) hence we can define $\psi = \tilde\psi_{E_0}$, $E_0 \in \mathcal{H}(V,\tilde V)$.
$E_0 \in \mathcal{H}(V,\tilde V)$.

We can now state the following

THEOREM. *Let* V *be a vector space of dimension* n *over* F *and let* $\tilde V$ *be a linear subspace of dimension* $m (n > m > 0)$. *There exists a canonical* W_F*-isomorphism* $\psi : D(V/\tilde V) \underset{W_F}{\otimes} P_{W_F}(V,\tilde V) \xrightarrow{\approx} D(V)$.

Remark. This tensor product decomposition corresponds in terms of dimensions to the factorization

$$(q-1)(q^2-1)\cdots(q^{n-1}-1) = (q-1)(q^2-1)\cdots(q^{n-m-1}-1)$$
$$\times (q^{n-m}-1)(q^{n-m+1}-1)\cdots(q^{n-1}-1)\,.$$

3.11 Applying Theorem 3.10 repeatedly we get the following

COROLLARY. *Let* V *be a vector space of dimension* n *over* F *and let* $\tilde V_{i_1} \subset \tilde V_{i_2} \subset \cdots \subset \tilde V_{i_k}$ *be a linear flag in* V $(\dim_F \tilde V_{i_a} = i_a)$ *with* $n > i_k > \cdots > i_2 > i_1 > 0$. *There exists a canonical* W_F*-isomorphism*

$$\psi : D(V/\tilde{V}_{i_k}) \underset{W_F}{\otimes} \left(\underset{1 \le a \le k}{\underset{W_F}{\otimes}} P_{W_F}(V/\tilde{V}_{i_{a-1}}, \tilde{V}_{i_a}/\tilde{V}_{i_{a-1}}) \right) \approx D(V)$$

where we use the convention $\tilde{V}_{i_0} = 0$.

This result describes the restriction of the GL(V)-module D(V) to a proper parabolic subgroup of GL(V) (the stabilizer of $\tilde{V}_{i_1} \subset \tilde{V}_{i_2} \subset \cdots \subset \tilde{V}_{i_k}$).

Take for example a complete linear flag $(\tilde{V}_1 \subset \tilde{V}_2 \subset \cdots \subset \tilde{V}_{n-1})$ in V. Then we have a canonical W_F-isomorphism

$$\psi : D(V/\tilde{V}_{n-1}) \underset{W_F}{\otimes} \left(\underset{1 \le a \le n-1}{\underset{W_F}{\otimes}} P_{W_F}(V/\tilde{V}_{a-1}, \tilde{V}_a/\tilde{V}_{a-1}) \right) \approx D(V)$$

describing the restriction of D(V) to a Borel subgroup of GL(V). In terms of dimensions, this corresponds to the factorization of $(q-1)(q^2-1)\cdots(q^{n-1}-1)$ into the factors $1, (q-1), (q^2-1), \cdots, (q^{n-1}-1)$.

3.12 Let $\tilde{V} \subset V$ be as in the Theorem 3.10. Let $U_{\tilde{V}}^{V}$ be the subgroup of all $t \in GL(V)$ such that $t|\tilde{V}$ = identity and $t|V/\tilde{V}$ = identity. This is clearly an elementary abelian p-group of order $q^{m(n-m)}$.

We have the following

PROPOSITION. *For any* $\tilde{V} \subset V$, $0 < \dim V < \dim \tilde{V}$, $\sum_{t \in U_{\tilde{V}}^{V}} t$ *acts as zero on* D(V).

Proof. We use the isomorphism ψ of 3.10 which clearly commutes with all $t \in G_{\tilde{V}}^{V}$ where $G_{\tilde{V}}^{V}$ is the maximal parabolic subgroup of GL(V) consisting of all $t : V \to V$ such that $t(\tilde{V}) = \tilde{V}$. ($D(V/\tilde{V})$ and $P_{W_F}(V,\tilde{V})$ can be regarded as $G_{\tilde{V}}^{V}$-modules in a natural way.) Note that any $t \in U_{\tilde{V}}^{V}$ acts as the identity on $D(V/\tilde{V})$. It is then sufficient to prove that $\sum_{t \in U_{\tilde{V}}^{V}} t$ acts as zero on $P_{W_F}(V, \tilde{V})$.

Let $u \in P_{W_F}(V, \tilde{V})$ and let $(V_{n-m} \subset V_{n-m+1} \subset \cdots \subset V_{n-1})$ be an incomplete flag in V such that $V_i + \tilde{V} = V$ $(n-m \leq i \leq n-1)$. We have

$$\left(\sum_{t \in U^V_{\tilde{V}}} tu\right)(V_{n-m} \subset V_{n-m+1} \subset \cdots \subset V_{n-1})$$

$$= \sum_{t \in U^V_{\tilde{V}}} u(t^{-1}V_{n-m} \subset t^{-1}V_{n-m+1} \subset \cdots \subset t^{-1}V_{n-1})$$

$$= \sum_{(V'_{n-m+1} \subset \cdots \subset V'_{n-1})} \quad \sum_{s \in U^{V_{n-m+1}}_{\tilde{V}}} u(s^{-1}V_{n-m} \subset V'_{n-m+1} \subset \cdots \subset V'_{n-1}) = 0$$

where $(V'_{n-m+1} \subset \cdots \subset V'_{n-1})$ runs over all distinct flags of the form $(t^{-1}V_{n-m+1} \subset \cdots \subset t^{-1}V_{n-1})$ for some $t \subset U^V_{\tilde{V}}$. The last equality follows from the cycle condition satisfied by u. The proposition is proved.

More generally, let $\tilde{\mathcal{V}} = (\tilde{V}_{i_1} \subset \tilde{V}_{i_2} \subset \cdots \subset \tilde{V}_{i_k})$ be a flag in V as in 3.11. Let $G^V_{\tilde{\mathcal{V}}}$ be the parabolic subgroup of $GL(V)$ consisting of all $t \in GL(V)$ such that $t(\tilde{V}_{i_a}) = \tilde{V}_{i_a}$ $(1 \leq a \leq k)$. Let $U^V_{\tilde{\mathcal{V}}}$ be the subgroup of $G^V_{\tilde{\mathcal{V}}}$ consisting of all $t \in G^V_{\tilde{\mathcal{V}}}$ such that $t|\tilde{V}_{i_a}/\tilde{V}_{i_{a-1}}$ = identity, $(1 \leq a \leq k+1)$ where $V_{i_0} = 0$, $V_{i_{k+1}} = V$. We have the following

COROLLARY. $\sum_{t \in U^V_{\tilde{\mathcal{V}}}} t$ *acts as zero on* $D(V)$.

Proof. This follows easily by induction on k from the proposition.

Remark. The conclusion of this corollary means that as a $GL(V)$-module, $D(V)$ satisfies the cusp-conditions hence it belongs to the discrete series (see [6]).

CHAPTER 4

THE CHARACTER OF D(V) AND THE EIGENVALUE $\lambda(V)$

4.1 Let V be a vector space of dimension $n \geq 1$ over a finite field F. Let $t: V \to V$ be an automorphism of V. We associate to t a Witt vector $\tilde{\mathrm{Tr}}(t|V) \in W_F$ as follows. Let F' be a finite extension field of F so that the eigenvalues $\lambda_1, \lambda_2, \cdots, \lambda_n$ of t lie in F'. We define $\tilde{\mathrm{Tr}}(t|V) = \tilde{\lambda}_1 + \tilde{\lambda}_2 + \cdots + \tilde{\lambda}_n \in W_{F'}$. It is clear that the set $(\lambda_1, \lambda_2, \cdots, \lambda_n)$ is invariant under $\mathrm{Gal}(F'/F)$. It follows that the set $(\tilde{\lambda}_1, \tilde{\lambda}_2, \cdots, \tilde{\lambda}_n)$ is also invariant under $\mathrm{Gal}(F'/F)$ (which acts on $W_{F'}$) hence $\tilde{\lambda}_1 + \tilde{\lambda}_2 + \cdots + \tilde{\lambda}_n$ is invariant under $\mathrm{Gal}(F'/F)$. It follows that $\tilde{\mathrm{Tr}}(t/V) \in W_F$ (see 3.6). It is easy to see that $\tilde{\mathrm{Tr}}(t|V)$ is independent of the choice of F'. Note that $\tilde{\mathrm{Tr}}(t|V)$ is precisely the *Brauer trace* of $t: V \to V$.

4.2 We say that an automorphism $t: V \to V$ is *anisotropic* if there is no subspace of V invariant under t other than 0 and V. In particular, if $\dim V = 1$, any $t: V \xrightarrow{\approx} V$ is anisotropic. In this section we shall prove the following

PROPOSITION. *If* $t: V \xrightarrow{\approx} V$ *is anisotropic then*

$$\mathrm{Tr}_{W_F}(t|DV) = (-1)^{n-1}\, \tilde{\mathrm{Tr}}(t|V)\,.$$

Proof. The proposition is obvious when $n = 1$. Assume now that $n \geq 2$. Let

$$M_1 = D(V) \oplus \Big(\bigoplus_{V_{n-2} \subset V} D(V_{n-2})\Big) \oplus \Big(\bigoplus_{V_{n-4} \subset V} D(V_{n-4})\Big) \oplus \cdots$$

and

$$M_2 = \Big(\bigoplus_{V_{n-1}\subset V} D(V_{n-1})\Big) \oplus \Big(\bigoplus_{V_{n-3}\subset V} D(V_{n-3})\Big) \oplus \cdots .$$

M_1 and M_2 are GL(V)-modules in a natural way. Since t has order prime to p (t is anisotropic) the eigenvalues of t on M_1 are in 1-1 correspondence with the eigenvalues of t on $M_1 \otimes_{W_F} F$ and similarly for M_2 and $M_2 \otimes_{W_F} F$.

Let $\mathcal{S}_i$ be the set of eigenvalues of t on $M_i \otimes F$ $(i = 1, 2)$ counted with multiplicity. From the exact sequence 1.13(c) it follows that

$$\mathcal{S}_1 \cup (\lambda_1, \lambda_2, \cdots, \lambda_n) = \mathcal{S}_2 \quad (n \text{ even})$$

$$\mathcal{S}_2 \cup (\lambda_1, \lambda_2, \cdots, \lambda_n) = \mathcal{S}_1 \quad (n \text{ odd})$$

where $(\lambda_1, \lambda_2, \cdots, \lambda_n)$ are the eigenvalues of t on V. Let $\tilde{\mathcal{S}}_i$ be the set of eigenvalues of t on $M_i (i = 1, 2)$. It follows that

$$\tilde{\mathcal{S}}_1 \cup (\tilde{\lambda}_1, \tilde{\lambda}_2, \cdots, \tilde{\lambda}_n) = \tilde{\mathcal{S}}_2 \quad (n \text{ even})$$

$$\tilde{\mathcal{S}}_2 \cup (\tilde{\lambda}_1, \tilde{\lambda}_2, \cdots, \tilde{\lambda}_n) = \tilde{\mathcal{S}}_1 \quad (n \text{ odd}) .$$

This implies that

$$\mathrm{Tr}_{W_F}(t|D(V)) = \sum_{1\le i\le n-1} (-1)^{n-1-i}\, \mathrm{Tr}_{W_F}\Big(t\Big|\bigoplus_{V_i\subset V} D(V_i)\Big) + (-1)^{n-1}\, \tilde{\mathrm{Tr}}(t|V) .$$

Next we observe that $\mathrm{Tr}_{W_F}(t|\bigoplus_{V_i\subset V} D(V_i)) = 0$ for $1 \le i \le n-1$ because t acts on the set of i-dimensional linear subspaces of V without any fixed point (since t is anisotropic). The proposition is proved.

4.3 We shall prove the following:

PROPOSITION. *Suppose we have* $\tilde{V} \subset V$ $(0 < \dim \tilde{V} = m < n)$ *and* $t : V \approx V$, $t(\tilde{V}) = \tilde{V}$ *such that* $t|\tilde{V}$ *is anisotropic. Then* $\mathrm{Tr}_{W_F}(t|P_{W_F}(V,\tilde{V})) = (-1)^{m-1}(N-1)$ *where* N *is the number of linear subspaces* $V_{n-m} \subset V$ *such that* $V_{n-m} + \tilde{V} = V$ *and* $t(V_{n-m}) = V_{n-m}$.

Proof. From the exact sequence 1.13(e) we see that

$$\mathrm{Tr}_{W_F}(t|P_{W_F}(V,\tilde{V})) = \sum_{n-m\leq i\leq n-i} (-1)^{n-1-i}\mathrm{Tr}_{W_F}(t|\bigoplus_{V_i \pitchfork \tilde{V}} P_{W_F}(V_i,V_i\cap\tilde{V})) + (-1)^{m-1}$$

The terms in the right hand side corresponding to $n-m+1 \leq i \leq n-1$ vanish since t acts on the set $\{V_i \mid V_i \subset V, V_i \pitchfork \tilde{V}\}$ without any fixed point, by the assumption that $t|V$ is anisotropic. On the other hand we have clearly

$$\mathrm{Tr}_{W_F}(t|\bigoplus_{V_{n-m} \pitchfork \tilde{V}} P_{W_F}(V_{n-m}, V_{n-m}\cap\tilde{V})) = N$$

where N is the number of fixed points of t acting on the set $\{V_{n-m} | V_{n-m} \subset V, V_{n-m} \pitchfork \tilde{V}\}$ and the proposition is proved.

4.4 In this section we shall give a formula for the number N of 4.3.

We say that $t: V \to V$ is *indecomposable* if whenever $V = V' \oplus V''$, $t(V') = V'$, $t(V'') = V''$, we must have $V' = 0$ or $V'' = 0$. In this case there is a unique subspace M of V such that $t(M) = M$ and $t|M$ is anisotropic.

In general, given $t: V \to V$ we can write V as the direct sum of t-invariant indecomposable subspaces which correspond to the Jordan cells of the matrix of t.

We have the following

PROPOSITION. *The hypotheses are the same as in* 4.3. *Write* $V/\tilde{V} = \bigoplus_{i\in I} V_{(i)}$ *where* $V_{(i)} \neq 0$, $t(V_{(i)}) = V_{(i)}$ *and* $t|V_{(i)}$ *is indecomposable for all* $i \in I$. *Let* $M_{(i)}$ *be the unique* t-*invariant subspace of* $V_{(i)}$ *such that* $t|M_{(i)}$ *is anisotropic. Let* J *be the set of all* $i \in I$ *such that there exists an isomorphism* $\tilde{V} \to M_{(i)}$ *commuting with* t. *Then either* $N = 0$ *or* $N = q^{me}$ *where* $e = \mathrm{card}(J)$.

Proof. Assume that $N > 0$. Then it follows from 4.3 that $N = \mathrm{card}[\mathrm{Hom}_t(\tilde{V}, V/\tilde{V})]$. Here Hom_t denotes the set of linear maps $\tilde{V} \to V/\tilde{V}$ commuting with t. We have

$$\mathrm{Hom}_t(\tilde{V}, V/\tilde{V}) = \bigoplus_{i \in I} \mathrm{Hom}_t(\tilde{V}, V_{(i)}) = \bigoplus_{i \in J} \mathrm{Hom}_t(\tilde{V}, M_{(i)}) .$$

It follows that $N = q^{se}$ where $s = \dim_F \mathrm{Hom}_t(\tilde{V}, \tilde{V})$. Since $t|\tilde{V}$ is anisotropic, $\mathrm{Hom}_t(\tilde{V}, \tilde{V})$ is a field with q^m elements hence $s = m$ and the proposition is proved.

Remark. We have $(-1)^{m-1}(N-1) = 0$ if and only if $N \neq 0$ and $e = 0$.

4.5 In this section we shall determine the character of the $GL(V)$-module $D(V)$ on an arbitrary element $t \in GL(V)$. By 4.2 we can assume that t is not anisotropic so that there exists some proper t-invariant subspace of V.

Choose a sequence $\tilde{V}_{i_1} \subset \tilde{V}_{i_2} \subset \cdots \subset \tilde{V}_{i_k}$ of linear subspaces of V, $\dim \tilde{V}_{i_a} = i_a$, $n > i_k > \cdots > i_2 > i_1 > 0$, $k \geq 1$, such that $t(\tilde{V}_{i_a}) = \tilde{V}_{i_a}$ $(1 \leq a \leq k)$ and t is anisotropic on

$$\tilde{V}_{i_1}, \tilde{V}_{i_2}/\tilde{V}_{i_1}, \cdots, \tilde{V}_{i_k}/\tilde{V}_{i_{k-1}}, V/\tilde{V}_{i_k} .$$

Using the isomorphism ψ of 3.11 and observing that ψ commutes with t we see that

$$\mathrm{Tr}_{W_F}(t|DV) = \mathrm{Tr}_{W_F}(t|D(V/\tilde{V}_{i_k})) \cdot \prod_{1 \leq a \leq k} \tau_a$$

where

$$\tau_a = \mathrm{Tr}_{W_F}(t|P_{W_F}(V/\tilde{V}_{i_{a-1}}, \tilde{V}_{i_a}/\tilde{V}_{i_{a-1}}) .$$

According to the remark in 4.4 we have $\tau_a = 0$ if and only if neither of the vector spaces $V/\tilde{V}_{i_k}, \tilde{V}_{i_k}/\tilde{V}_{i_{k-1}}, \cdots, \tilde{V}_{i_{a+1}}/\tilde{V}_{i_a}$ is isomorphic to $\tilde{V}_{i_a}/\tilde{V}_{i_{a-1}}$ by an isomorphism commuting with t. It follows that

$\prod_{1 \leq a \leq k} \tau_a \neq 0$ if and only if all vector spaces $\tilde{V}_{i_1}, \tilde{V}_{i_2}/\tilde{V}_{i_1}, \cdots, \tilde{V}_{i_k}/\tilde{V}_{i_{k-1}}$, $\tilde{V}/\tilde{V}_{i_k}$ are mutually isomorphic by isomorphisms commuting with t; in this case, or in case t is anisotropic we say that t is isotypic. We now assume that t is isotypic. Let $V = V_{(1)} \oplus V_{(2)} \oplus \cdots \oplus V_{(j)}$ be a decomposition of V such that $V_{(i)} \neq 0$, $V_{(i)}$ is t-invariant and $t|V_{(i)}$ is indecomposable $(1 \leq i \leq j)$. We can assume that

$$V_{(1)} = V_{i_{a(1)}}, V_{(1)} \oplus V_{(2)} = V_{i_{a(2)}}, \cdots, V_{(1)} \oplus V_{(2)} \oplus \cdots \oplus V_{(j-1)}$$
$$= V_{i_{a(j-1)}} \quad 1 \leq a(1) < a(2) < \cdots < a(j-1) \leq k \ .$$

Let $m = \dim_F \tilde{V}_{i_1} = \dim_F \tilde{V}_{i_2}/\tilde{V}_{i_1} = \cdots = \dim_F V/\tilde{V}_{i_k}$. From 4.3 and 4.4 it follows that $\tau_a = (-1)^{m-1}(-1) = (-1)^m$ for all a, $1 \leq a \leq k$, $a \neq a(h)$, $1 \leq h \leq j-1$ and that $\tau_a = (-1)^{m-1}(q^{m(j-h)}-1)$ if $a = a(h)$, $1 \leq h \leq j-1$. Then the product of the τ_a's is

$$\prod_{1 \leq a \leq k} \tau_a = (-1)^{mk+j-1}(q^m-1)(q^{2m}-1)\cdots(q^{(j-1)m}-1) \ .$$

On the other hand, according to 4.2,

$$\mathrm{Tr}_{W_F}(t|D(V/\tilde{V}_{i_k})) = (-1)^{\dim(V/\tilde{V}_{i_k})-1}\,\tilde{\mathrm{Tr}}(t|V/\tilde{V}_{i_k}) = (-1)^{m-1}\,\tilde{\mathrm{Tr}}(t|\tilde{V}_{i_1}) \ .$$

It follows that

$$\mathrm{Tr}_{W_F}(t|D(V/\tilde{V}_{i_k}))\cdot \prod_{1 \leq a \leq k} \tau_a = (-1)^{m(k+1)+j}\tilde{\mathrm{Tr}}(t|\tilde{V}_{i_1})(q^m-1)(q^{2m}-1)\cdots(q^{(j-1)m}-1)$$

Note that $m(k+1) = n$.

We can now state the following

THEOREM. *Let* $t : V \overset{\approx}{\to} V$ *be an automorphism.* ($\dim V = n \geq 1$). *Let* j *be the number of terms in a decomposition of* V *into non-zero, indecomposable,* t*-invariant subspaces. Let* $\tilde{V}_{an}$ *be some* t*-invariant subspace of* V *such that* $\tilde{V}_{an} \neq 0$ *and* $t|\tilde{V}_{an}$ *is anisotropic,* ($\dim \tilde{V}_{an} = m$). *Then*

$$\mathrm{Tr}_{W_F}(t|DV) = (-1)^{n+j}\, \tilde{\mathrm{Tr}}(t|\tilde{V}_{an})(q^m-1)(q^{2m}-1)\cdots(q^{(j-1)m}-1)$$

if t *is isotypic and*

$$\mathrm{Tr}_{W_F}(t|DV) = 0 \quad \textit{otherwise.}$$

Remarks:

1. In case $j = 1$, i.e., $t : V \to V$ indecomposable, we regard $(q^m-1)(q^{2m}-1)\cdots(q^{(j-1)m}-1)$ as being equal to 1. In this case the formula reduces to

$$\mathrm{Tr}_{W_F}(t|DV) = (-1)^{n+1}\, \tilde{\mathrm{Tr}}(t|\tilde{V}_{an})\,.$$

2. The theorem identifies the character $t \to \mathrm{Tr}_{W_F}(t|DV)$ with one in Green's classification of all the irreducible characters of $GL(V)$. (cf. [5].)

4.6 Our aim now is to compute the Witt vector $\lambda(V)$ introduced in 3.9. We recall that $\lambda(V) \in W_F$ is defined by the identity $Tf = \lambda(V)\cdot f$, $f \in D(V)$. Here we regard $D(V)$ as a W_F-submodule of $\mathcal{F}_{W_F}$, the set of all functions $f : Y \to W_F$ (see 3.1). $GL(V)$ acts on $\mathcal{F}_{W_F}$ by $t(E_0 \subset E_1 \subset \cdots \subset E_{n-1}) = (t^{-1}(E_0) \subset t^{-1}(E_1) \subset \cdots \subset t^{-1}(E_{n-1}))$; this corresponds to the natural transitive action of $GL(V)$ on Y. Let $\varepsilon = (E_0 \subset E_1 \subset \cdots \subset E_{n-1}) \in Y$ be a fixed flag, and let B_ε be the group of all $t \in GL(V)$ such that $t(E_i) = (E_i)$ $(0 \leq i \leq n-1)$; this is also the group of all $t \in \mathrm{Aff}(E_{n-1})$ such that $t(E_i) = (E_i)$ $0 \leq i \leq n-1$.

Let Ω be an algebraically closed field of characteristic zero. We have the following

LEMMA. *Let* G *be a finite group and let* B *be a subgroup of* G. *Let* $\mathcal{F}$ *be the* Ω*-vector space of all functions* $f: G \to \Omega$ *such that* $f(gb) = f(g)$, $b \in B$, $g \in G$. *Then* $\mathcal{F}$ *is a* G*-module under* $(gf)(g') = f(g^{-1}g')$, $g, g' \in G$, $f \in \mathcal{F}$. *Let* $M \subset \mathcal{F}$ *be a* G*-invariant subspace of* $\mathcal{F}$. *Assume that up to a non-zero scalar there is a unique* $f_0 \in M$ *such that* $bf_0 = f_0$, $b \in B$. *Then we can take* $f_0(g) = \sum_{b \in B} \mathrm{Tr}_\Omega(bg^{-1}|M)$.

The proof is straightforward and will be omitted.

We apply the lemma to the case $G = GL(V)$, $B = B_\varepsilon$, $M = D(V) \otimes_{W_F} \Omega$ where the tensor product is with respect to some homomorphism $W_F \to \Omega$. The hypothesis of the lemma is then satisfied (cf. 2.1 and 3.8). It follows that the function $f_0 : Y \to W_F$ defined by $f_0(g\varepsilon) = \sum_{b \in B} \mathrm{Tr}_{W_F}(bg^{-1}|D(V))$ is the unique function $Y \to W_F$ such that $f_0(\varepsilon) = |B_\varepsilon|$, $f_0(b\varepsilon') = f_0(\varepsilon')$, $\varepsilon' \in Y$, $b \in B_\varepsilon$ and $f_0 \in D(V)$. Applying T to f_0 we get $Tf_0 = \lambda(V) f_0$, since $f_0 \in D(V)$. Evaluating at $\varepsilon \in Y$ we get $\lambda(V) = \frac{1}{|B_\varepsilon|}(Tf_0)(\varepsilon)$.

4.7 We recall that $(Tf_0)(\varepsilon) = (-1)^{n-1} \sum_{\varepsilon' \in \Theta(\varepsilon)} f_0(\varepsilon')$ where $\Theta(\varepsilon')$ is the set of all flags $\varepsilon' = (E'_0 \subset E'_1 \subset \cdots \subset E'_{n-1})$ such that $E'_0 \in E_{n-1} \setminus E_{n-2}$, $E'_1 \| [E_0]$, $E'_2 \| [E_1]$, $\cdots$, $E'_{n-1} \| [E_{n-2}]$. Let $\mathcal{R}$ be the set of all $t \in GL(V)$ such that $t\varepsilon \in \Theta(\varepsilon)$. Any $t \in \mathcal{R}$ gives rise to a basis $\tilde{\beta}(t)$ of V defined by $\tilde{\beta}(t) = (E_0, t^{-1}E_0, t^{-2}E_0, \cdots, t^{-(n-1)}E_0)$. Let $(V_1 \subset V_2 \subset \cdots \subset V_{n-1})$ be the complete linear flag in V defined by $V_1 \| E_1, V_2 \| E_2, \cdots, V_{n-1} \| E_{n-1}$.

A basis $\beta = (v_0, v_1, \cdots, v_{n-1})$ of V is said to be *adapted* to ε if $v_0 = E_0$, $v_1 \in V_1 \setminus 0$, $v_2 \in V_2 \quad V_1, \cdots, v_{n-1} \in V_{n-1} \setminus V_{n-2}$.

Let $\mathcal{B}$ be the set of all basis of V adapted to ε. Then B_ε operates freely and transitively on $\mathcal{B}$ by $b(v_0, v_1, \cdots, v_{n-1}) = (b(v_0), b(v_1), \cdots, b(v_{n-1}))$. It is clear that for any $t \in \mathcal{R}$ we have $\tilde{\beta}(t) \in \mathcal{B}$. Given $\beta \in \mathcal{B}$, $\beta = (v_0, v_1, \cdots, v_{n-1})$, define $\mathcal{R}_\beta = \{t \in \mathcal{R} | t(v_1) = v_0, t(v_2) =$

$v_1, \cdots, t(v_{n-1}) = v_{n-2}\}$. It is clear that $t \in \mathcal{R}$ if and only if $\tilde{\beta}(t) = \beta$. We have the following

LEMMA. (i) *For any* $b \in B_\varepsilon$ *and* $\beta \in \mathcal{B}$ *we have* $b\mathcal{R}_\beta b^{-1} = \mathcal{R}_{b\beta}$.

(ii) *For any* $\beta \in \mathcal{B}$, *and any* $t \in \mathcal{R}$ *there exist unique elements* $t' \in \mathcal{R}_\beta$ *and* $b \in B_\varepsilon$ *such that* $t = t'b$.

Proof. Let $t \in \mathcal{R}$; we have $\tilde{\beta}(btb^{-1}) = (E_0, bt^{-1}b^{-1}(E_0), bt^{-2}b^{-1}(E_0), \cdots, bt^{-(n-1)}b^{-1}(E_0)) = (E_0, bt^{-1}(E_0), \cdots, bt^{(n-1)}(E_0)) = b\tilde{\beta}(t)$ and (i) follows. It is easy to see that for $\beta = (v_0, v_1, \cdots, v_{n-1}) \in \mathcal{B}$ and $t \in \mathcal{R}$ we have $\beta' = (v_0, t^{-1}(v_0), t^{-1}(v_1), \cdots, t^{-1}(v_{n-2})) \in \mathcal{B}$. Since B_ε acts transitively on $\mathcal{B}$, there exists $b \in B_\varepsilon$ such that $b\beta = \beta'$. Then we have $bv_0 = v_0$, $bv_i = t^{-1}(v_{i-1})$ $(1 \le i \le n-1)$. Hence $tb(v_0) = t(v_0)$ and $tbv_i = v_{i-1}$ $(1 \le i \le n-1)$, i.e., $tb \in \mathcal{R}_\beta$. It follows that $t \in \mathcal{R}_\beta \cdot b^{-1}$ and the existence part of (ii) is proved.

Assume now that $tb = t'b'$, $t, t' \in \mathcal{R}_\beta$, $b, b' \in B_\varepsilon$; we wish to prove that $b = b'$. We can assume that $b' = 1$ hence $tb = t'$. We have $t(v_i) = v_{i-1}$, $t'(v_i) = v_{i-1}$ $(1 \le i \le n-1)$ hence $tbv_i = v_{i-1}$. It follows that $v_i = t^{-1}v_{i-1}$, $bv_i = t^{-1}v_{i-1}$ $(1 \le i \le n-1)$ hence $bv_i = v_i$ $(1 \le i \le n-1)$. Since clearly $tv_0 = v_0$ it follows $b = 1$ and the lemma is proved.

We fix now some $\beta_0 \in \mathcal{B}$. Given $\varepsilon' \in \Theta(\varepsilon)$ we have $\varepsilon' = t(\varepsilon)$ for some $t \in \mathcal{R}$. According to the lemma (ii) we have $t = t' \cdot b$ where $t' \in \mathcal{R}_{\beta_0}$, $b \in B_\varepsilon$; hence $\varepsilon' = t'b\varepsilon = t'\varepsilon$. It is clear that if $t', t'' \in \mathcal{R}_{\beta_0}$ and $t' \neq t''$, then $t'(\varepsilon) \neq t''(\varepsilon)$. In fact, otherwise we would have $t' = t'' \cdot b$, $b \in B_\varepsilon$, $b \neq 1$ which contradicts the unicity part of lemma (ii).

It follows that the elements of $\Theta(\varepsilon)$ are precisely the flags

$$\{t'(\varepsilon) \mid t' \in \mathcal{R}_{\beta_0}\} .$$

We can then write (cf. 4.6)

$$\lambda(V) = (-1)^{n-1} \frac{1}{|B_\varepsilon|} \sum_{t' \in \mathcal{R}_{\beta_0}} f_0(t'\varepsilon) = (-1)^{n-1} \frac{1}{|B_\varepsilon|} \sum_{t' \in \mathcal{R}_{\beta_0}} \sum_{b \in B_\varepsilon} \mathrm{Tr}_{W_F}(bt'^{-1} | D(V)).$$

According to the lemma (ii) we can replace the sum over $t' \in \mathcal{R}_{\beta_0}$ and $b \in B_\varepsilon$ by a sum over $t \in \mathcal{R}$:

$$\lambda(V) = (-1)^{n-1} \frac{1}{|B_\varepsilon|} \sum_{t \in \mathcal{R}} \mathrm{Tr}_{W_F}(t^{-1} \mid D(V)) .$$

Using the lemma (i) we have

$$\lambda(V) = (-1)^{n-1} \frac{1}{|B_\varepsilon|} \sum_{\substack{t' \in \mathcal{R}_{\beta_0} \\ b \in B_\varepsilon}} \mathrm{Tr}_{W_F}(bt'^{-1}b^{-1} | D(V))$$

$$= (-1)^{n-1} \sum_{t' \in \mathcal{R}_{\beta_0}} \mathrm{Tr}_{W_F}(t'^{-1} | D(V)) .$$

4.8 Consider the action of $t' \in \mathcal{R}_{\beta_0}$ on the basis $\beta_0 = (v_0, v_1, \cdots, v_{n-1})$ of V. We have $tv_0 \in E_{n-1} \setminus E_{n-2}$ hence $tv_0 = v_0 + a_1 v_1 + a_2 v_2 + \cdots + a_{n-2} v_{n-2} + a_{n-1} v_{n-1}$, $a_i \in F$, $1 \leq i \leq n-1$, $a_{n-1} \neq 0$

$$tv_1 = v_0, \ tv_2 = v_1, \ \cdots, \ tv_{n-1} = v_{n-2} .$$

Hence t' has the matrix

$$\begin{pmatrix} 1 & a_1 & a_2 & \cdots & a_{n-1} \\ 1 & 0 & \cdots & \cdots & 0 \\ 0 & 1 & \ddots & & \vdots \\ \vdots & \ddots & \ddots & \ddots & \vdots \\ 0 & \cdots & 0 & 1 & 0 \end{pmatrix}$$

with respect to $\beta_0 (a_{n-1} \neq 0)$. Conversely any $t' \in GL(V)$ which has such a matrix with respect to β_0, belongs to $\mathcal{R}_{\beta_0}$. It follows that $\mathcal{R}_{\beta_0}$ consists of regular elements in GL(V) of trace 1 and in fact $\mathcal{R}_{\beta_0}$ contains one element in each conjugacy class of regular elements of trace 1 in GL(V). (We recall that an element $t \in GL(V)$ is called *regular* (in the sense of Steinberg) if its centralizer is abelian (cf. [16].) It is easy to see that $t \in GL(V)$ is regular and isotypic (cf. 4.5) if and only if t is indecomposable (cf. 4.4). Since $\mathrm{tr}_{W_F}(t'^{-1}|DV) = 0$ if t'^{-1} is not isotypic we have (cf. 4.7)

$$\lambda(V) = (-1)^{n-1} \sum \mathrm{Tr}_{W_F}(t'^{-1}|D(V))$$

where the sum is over a cross-section of the conjugacy classes of indecomposable elements $t' \in GL(V)$ of trace 1. These are clearly in 1–1 correspondence with the set of orbits of $\mathrm{Gal}(F'/F)$ on the set $\{x \in F' | \mathrm{Tr}_{F'/F}(x) = 1\}$ where F' is a fixed extension field of degree n of F. Using Remark 1 in 4.5 we have

$$\lambda(V) = (-1)^{n-1} \sum (-1)^{n-1} \mathrm{Tr}(t'^{-1}|V) = \sum (\tilde{x}^{-1} + \tilde{x}^{-q} + \cdots + \tilde{x}^{-q^{d-1}})$$

where the sum is over all orbits $(x, x^q, \cdots, x^{q^{d-1}})$ of $\mathrm{Gal}(F'/F)$ on the set $\{x \in F' | \mathrm{Tr}_{F'/F}(x) = 1\}$.

We have proved the following

THEOREM.
$$\lambda(V) = \sum_{\substack{x \in F' \\ \mathrm{Tr}_{F'/F} x = 1}} \tilde{x}^{-1} \in W_F .$$

Remark. The terms of the sum defining $\lambda(V)$ lie in $W_{F'}$, but the sum is invariant under $\mathrm{Gal}(F'/F)$ hence lies in W_F. It is clear that $\lambda(V)$ is independent of the choice of F'.

4.9 We know that $\lambda(V) \equiv 1 \pmod p$ (cf. 3.9). It follows then from Theorem 4.8 that

$$\sum_{\substack{x \in F' \\ \mathrm{Tr}_{F'/F}(x) = 1}} x^{-1} = 1 .$$

This identity can be also proved directly as follows.

Let

$$a = \sum_{\substack{x \in F' \\ Tr_{F'/F}(x)=1}} x^{-1},$$

and

$$b = \sum_{x \in F'^*} x^{-1}\, Tr_{F'/F}(x)\,.$$

Note that $(yx)^{-1}\, Tr_{F'/F}(yx) = x^{-1}\, Tr_{F'/F}(x)$ for $y \in F^*$, $x \in F'^*$. It follows that $b = (q-1)a = -a$. We have

$$b = \sum_{x \in F'^*} x^{-1}(x + x^q + \cdots + x^{q^{n-1}})$$

$$= \sum_{x \in F'^*} 1 + \sum_{x \in F'^*} x^{q-1} + \cdots + \sum_{x \in F'^*} x^{q^{n-1}-1} = (q^n-1) = -1\,.$$

It follows that $a = 1$.

4.10 We shall now determine the smallest field over which the $GL(V)$-module $D(V)$ can be defined. Let $K_{F,1}$ be the subfield of the quotient field Q_F of W_F generated by the elements $\tilde{\lambda}(\lambda \in F^*)$. This is then isomorphic to the cyclotomic field $Q(\sqrt[q-1]{1})$. Let $K_{F,n}$ $(n \geq 2)$ be the subfield of Q_F generated by the elements $\tilde{\lambda}(\lambda \in F^*)$ and by $\lambda(V)$ where V is some F-vector space of dimension n. It is convenient to define $\lambda(V) = 1$ when $\dim_F V = 1$. We have, for example $K_{F,2} = Q(\sqrt{-2})$ when F is the field with 3 elements (in this case $\lambda(V) = \sqrt{-2}-1$, where $\sqrt{-2} \equiv -1 \pmod 3$)). Let $\mathcal{O}_{F,n} = K_{F,n} \cap W_F \subset Q_F$ $(n \geq 1)$. Then $\mathcal{O}_{F,n}$ is a discrete valuation ring in $K_{F,n}$ with a unique non-zero prime ideal $(= p \cdot \mathcal{O}_{F,n})$, with residue field F. Let V be an F-vector space of dimension 1 and let $\mathcal{O} = \mathcal{O}_{F,1}$. Define

$$D(V)_{\mathcal{O}} = \{f : V\backslash 0 \to \mathcal{O} \mid f(\lambda v) = \tilde{\lambda}^{-1} f(v), \lambda \in F^*, v \in V\backslash 0\}.$$

Then $D(V) = D(V)_{\mathcal{O}} \otimes_{\mathcal{O}} W_F$.

Let now V be an F-vector space of dimension $n \geq 2$. Consider the free $\mathcal{O} = \mathcal{O}_{F,n}$-module $\dot{\Delta}_{\mathcal{O}}(V)$ cf. (1.13). Let $\dot{\Delta}_{\mathcal{O}}(V)^{(k)}$, $k \in \mathbf{Z}/(k-1)\mathbf{Z}$ be the set of all $f \in \dot{\Delta}_{\mathcal{O}}(V)$ satisfying the homogeneity condition

$$f(\lambda E_0 \subset \lambda E_1 \subset \cdots \subset \lambda E_{n-1}) = \tilde{\lambda}^k f(E_0 \subset E_1 \subset \cdots \subset E_{n-1})$$

for all $\lambda \in F^*$ and $(E_0 \subset E_1 \subset \cdots \subset E_{n-1}) \in Y$. (Note that $\tilde{\lambda} \in \mathcal{O}$.) It is easy to see that $\dot{\Delta}_{\mathcal{O}}(V) = \bigoplus_{k \in \mathbf{Z}/(q-1)\mathbf{Z}} \dot{\Delta}_{\mathcal{O}}(V)^{(k)}$. Define

$$D(V)_{\mathcal{O}} = \{f \in \dot{\Delta}_{\mathcal{O}}(V)^{(-1)} \mid Tf = \lambda(V) f\}$$

and

$$D(V)_{\mathcal{O}}^{\perp} = \text{Image}\,(T - \lambda(V) : \Delta_{\mathcal{O}}(V)^{(-1)} \to \dot{\Delta}_{\mathcal{O}}(V)^{(-1)}) \ .$$

We have $\Delta_{\mathcal{O}}(V)^{(-1)} = D(V)_{\mathcal{O}} \oplus D(V)_{\mathcal{O}}^{\perp}$ and $D(V) \cong D(V)_{\mathcal{O}} \otimes_{\mathcal{O}} W_F$, $D(V)^{\perp} \cong D(V)_{\mathcal{O}}^{\perp} \otimes_{\mathcal{O}} W_F$ (this follows easily from 3.9). This shows that $D(V)$ can be defined as a $GL(V)$-module over $\mathcal{O} = \mathcal{O}_{F,n}$. It follows that $D(V)$ can be also defined over $K_{F,n}$. We can define $D(V)_{K_{F,n}} = D(V)_{\mathcal{O}} \otimes_{\mathcal{O}} K_{F,n}$. We now show that $K_{F,n}$ is the smallest field of characteristic zero over which $D(V)$ can be defined. For this it is enough to check that the field generated by the values of the character of $D(V)$ contains $K_{F,n}$. We have clearly $\tilde{\lambda} = (\text{rank}_{W_F}(DV))^{-1}\, \text{Tr}_{W_F}(\lambda \cdot 1 | D(V))$ where $\lambda \cdot 1$ denotes the automorphism $v \to \lambda v$ of $V, (\lambda \in F^*)$. Moreover from 4.8 it follows that $\lambda(V)$ is a $\mathbf{Z}$-linear combination of values of the character of $D(V)$ on regular elements in $GL(V)$. This proves our assertion.

We can state the following

PROPOSITION. *The field of definition of the* $GL(V)$*-module* $D(V)$ *coincides with the field generated by the values of its character and with the field* $K_{F,n}$ *generated by* $\tilde{\lambda}(\lambda \in F^*)$ *and* $\lambda(V)$.

4.11 In this section we shall study the algebraic number field $K_{F,n}$ defined in 4.10.

$K_{F,1}$ is clearly isomorphic with the cyclotomic field $Q(\sqrt[q-1]{1})$. Let F' be a field extension of degree n of F. Then the group $\Gamma = Gal(F'/F)$ acts on $Q_{F'}$ and leaves $K_{F',1}$ invariant.

PROPOSITION. $$K_{F,n} = K_{F',1}^{\Gamma} .$$

Proof. The inclusion $K_{F,n} \subset K_{F',1}^{\Gamma}$ is obvious. To prove the reverse inclusion, note that the subfield $K_{F',1}^{\Gamma}$ of Γ-invariants in $K_{F',1}$ is generated by elements of the form $\tilde{x} + \tilde{x}^q + \cdots + \tilde{x}^{q^{d-1}}$ where $(x, x^q, \cdots, x^{q^{d-1}})$ is any orbit of Γ on F'^*. This is precisely the field generated by the character of $D(V)$ where V is an F-vector space of dimension n. This is the same as $K_{F,n}$, cf. 4.10. The proposition is proved.

Remark. This proposition can be interpreted in the following way: Consider the field extension $K_{F,1} \subset K_{F',1}^{\Gamma}$ where $K_{F,1} \cong Q(\sqrt[q-1]{1})$ and $K_{F',1} \cong Q(\sqrt[q^n-1]{1})$. Then $\lambda(V)$ is a primitive element for this extension. Although this is purely a field-theoretic statement our proof is via the representation theory of $GL_n(F)$.

COROLLARY. $[K_{F,n} : Q] = \frac{1}{n} \cdot \phi(q^n-1)$ *where* ϕ *is the Euler function.*

This follows from well-known formula of Gauss: $[Q\sqrt[q^n-1]{1} : Q] = \phi(q^n-1)$ and from the fact that $|\Gamma| = n$.

4.12 In this section we shall show how from the distinguished member of the discrete series of $GL(V)$ one can obtain other members. We shall use the notation $K = K_{F,n}$. Since $\dot{\Delta}_K(V) = \dot{\Delta}_Q(V) \underset{Q}{\otimes} K$, the Galois group $Gal(K/Q)$ acts naturally on $\dot{\Delta}_K(V)$ by the formula $\gamma(f \otimes x) = f \otimes \lambda x$,

$f \in \dot{\Delta}_Q(V)$, $x \in K$, $\gamma \in \mathrm{Gal}(K/Q)$. It is clear that this action of $\mathrm{Gal}(K/Q)$ commutes with the action of $GL(V)$ and with the endomorphism T of $\dot{\Delta}_K(V)$. (Note that T is defined over Q.) Let $D(V)_K^{\gamma} = \gamma(D(V)_K)$. Then $D(V)_K^{\gamma}$ must be a K-linear subspace of $\dot{\Delta}_K(V)$ invariant under $GL(V)$ and under T. In fact, we can describe $D(V)_K^{\gamma}$ as the set of all $f \in \dot{\Delta}_K(V)$ such that

$$f(\lambda E_0 \subset \lambda E_1 \subset \cdots \subset \lambda E_{n-1}) = \gamma(\tilde{\lambda})^{-1} f(E_0 \subset E_1 \subset \cdots \subset E_{n-1})$$

for $\lambda \in F^*$, $(E_0 \subset E_1 \subset \cdots \subset E_{n-1}) \in Y$ and $Tf = \gamma(\lambda(V))f$. The $GL(V)$-module $D(V)_K^{\gamma}$ is in the discrete series of $GL(V)$, for any $\gamma \in \mathrm{Gal}(K/Q)$. In fact, by definition, an irreducible $GL(V)$-module D is in the discrete series if for any proper parabolic subgroup $G_{\mathcal{Q}}^V$ in $GL(V)$ we have the cusp condition $\sum_{t \in U_{\mathcal{Q}}^V} t = 0$ on D (see 3.12). Since the cusp conditions are defined over $\mathbf{Z}$, they are preserved by Galois automorphisms and our assertion follows. Note however that the $GL(V)$-modules $D(V)_K^{\gamma}$ do not in general exhaust the discrete series of $GL(V)$. They provide $\frac{\phi(q^n-1)}{n}$ members of the discrete series (cf. Corollary 4.11) while the total number of members in the discrete series is $\frac{q^n}{n} + 0(q^{n-1})$ (cf. [5], [13]).

CHAPTER 5

THE BRAUER LIFTING

5.1 Let G be a finite group and let A be some principal ideal domain. Let $M_A(G)$ (resp. $M'_A(G)$) be the category of all finitely generated free (resp. finitely generated) A-modules with a G-action which is A-linear; the maps are A-linear maps commuting with G. Let $R_A(G)$ (resp. $R'_A(G)$) denote the Grothendieck group of $M_A(G)$ (resp. $M'_A(G)$). This is an abelian group generated by the isomorphism classes of objects in $M_A(G)$ (resp. $M'_A(G)$) with one relation for every short exact sequence in $M_A(G)$ (resp. $M'_A(G)$). It is well known, and easy to prove, that the natural map $\alpha : R_A(G) \to R'_A(G)$ is an isomorphism. An inverse to α is obtained by associating to $M \in M'_A(G)$ the formal difference $L - L' \in R_A(G)$ where $0 \to L' \to L \to M \to 0$ is an exact sequence in $M'_A(G)$ with $L, L' \in M_A(G)$.

5.2 Let V be a vector space of dimension $n \geq 1$ over F. Consider the virtual representation

$$br(V) = (-1)^{n-1} D(V) + (-1)^{n-2} \bigoplus_{V_{n-1} \subset V} D(V_{n-1}) + \cdots + (-1)^0 \bigoplus_{V_1 \subset V} D(V_1)$$

of $GL(V)$. Note that $GL(V)$ operates naturally on the free W_F-module $\bigoplus_{V_i \subset V} D(V_i)$. When $1 \leq i \leq n-1$ this can be regarded as the representation of $GL(V)$ induced by a representation of the maximal parabolic subgroup $G^V_{V_i}$, which is trivial on $U^V_{V_i}$, for some fixed $V_i \subset V$; see 3.12 for the definition of $G^V_{V_i}$ and $U^V_{V_i}$. $br(V)$ can be considered as an element in $R_{W_F}(GL(V))$. Under the natural map $R_{W_F}(GL(V)) \to R_F(GL(V))$, $br(V)$

goes to V, regarded as a GL(V)-module in the natural way. (This follows from 3.8 and the exact sequence 1.13(c).)

We shall now study the behavior of br(V) under extensions.

Let $\tilde{V} \subset V$ be an m-dimensional linear subspace of V, $(0 < m < n)$. Let M be any free W_F-module with a GL(V)-action. Let $M' = \{f \in M \mid \left(\sum_{t \in U_{\tilde{V}}^V} t\right) f = 0\}$ and $M'' = M/M'$. Then M' and M'' are $G_{\tilde{V}}^V$-modules since $U_{\tilde{V}}^V$ is a normal subgroup of $G_{\tilde{V}}^V$. It is obvious that the exact sequence of W_F-modules $0 \to M' \to M \to M'' \to 0$ is split over W_F but the splitting is in general not compatible with the action of $G_{\tilde{V}}^V$. For example if $M = D(V)$, we have $M' = M = D(V/\tilde{V}) \underset{W_F}{\otimes} P_{W_F}(V, \tilde{V})$, by Proposition 3.12 (cusp condition) and 3.10. It follows that $M'' = 0$ in this case.

PROPOSITION. *Let* $M = \bigoplus_{V_i \subset V} D(V_i)$ $(0 < i < n)$. *Then*

$$\text{(i)}\quad M' = \bigoplus_{\substack{V_s \subset V \\ \tilde{V} \subset V_s \\ m < s \leq m+i}} D(V_s/\tilde{V}) \underset{W_F}{\otimes} Q_i(V_s, \tilde{V})$$

and

$$\text{(ii)}\quad M'' = \bigoplus_{V_i \subset \tilde{V}} D(V_i) \oplus \left(\bigoplus_{V_i \subset V/\tilde{V}} D(V_i) \right)$$

where $Q_i(V_s, \tilde{V}) = \bigoplus_{\substack{V_i \subset V_s \\ V_i + \tilde{V} = V_s}} P_{W_F}(V_i, V_i \cap \tilde{V})$ *if* $m < s < m+i$,

$$Q_i(V_s, \tilde{V}) = \ker \left(\bigoplus_{\substack{V_i \subset V_s \\ V_i + \tilde{V} = V_s}} P_{W_F}(V_i, 0) \to W_F \right) \quad \textit{if } s = m + i$$

and

$$Q_i(V_s, \tilde{V}) = 0, \quad \textit{otherwise}.$$

Proof. Let $f = (f^{(\alpha)}) \in \bigoplus_{V_i \subset V} D(V_i)$, $f^{(\alpha)} \in D(V_i^{(\alpha)})$ where α parametrizes the set of i-dimensional linear subspaces of V. Assume that $\left(\sum_{t \in U_{\tilde{V}}^V} t\right) f = 0$.

Let $\overline{f}^{(\alpha)} = (0 \cdots 0, f^{(\alpha)}, 0 \cdots 0)$ be the element obtained from f by replacing all the coordinates (except for the one on place α) by zero. Assume that $V_i \cap \tilde{V} \neq 0, \tilde{V}$, where $V_i = V_i^{(\alpha)}$. Write $U_{\tilde{V}}^V$ as a direct sum $U' \oplus U_{V_i \cap \tilde{V}}^{V_i}$ (note that $U_{V_i \cap \tilde{V}}^{V_i} = \{t \in U_{\tilde{V}}^V \mid t(V_i) = V_i\}$. Then we have

$$\left(\sum_{t \in U_{\tilde{V}}^V} t\right) \overline{f}^{(\alpha)} = \left(\sum_{t \in U'} t\right)\left(\sum_{t \in U_{V_i \cap \tilde{V}}^{V_i}} t\right) \overline{f}^{(\alpha)} = 0$$

(cf. Proposition 3.12).

Assume now that $V_i \subset \tilde{V}$, where $V_i = V_i^{(\alpha)}$. Then clearly $t\overline{f}^{(\alpha)} = \overline{f}^{(\alpha)}$ for all $t \in U_{\tilde{V}}^V$, hence $\left(\sum_{t \in U_{\tilde{V}}^V} t\right) \overline{f}^{(\alpha)} = |U_V^V| \cdot \overline{f}^{(\alpha)}$. Let now $V_s \subset \tilde{V}$ be such that $s = m + i$. Let $\mathfrak{F}(V_s)$ be the set of all i-dimensional linear subspaces V_i of V_s such that $V_i + \tilde{V} = V_s$. In this case, $D(V_i) = D(V_s/\tilde{V})$ for all $V_i \subset \mathfrak{F}(V_s)$. Hence a family $(f^{(\alpha)})_{\alpha \in \mathfrak{F}(V_s)}$, $f^{(\alpha)} \in D(V_i^{(\alpha)})$ is the same as a family $(\tilde{f}^{(\alpha)})_{\alpha \in \mathfrak{F}(V_s)}$, $\tilde{f}^{(\alpha)} \in D(V_s/\tilde{V})$.

The group $U_{\tilde{V}}^V$ acts on $\mathfrak{F}(V_s)$ by permutations (it leaves V_s invariant), and it is clear that $\left(\sum_{t \in U_{\tilde{V}}^V} t\right)(f^{(\alpha)})_{\alpha \in \mathfrak{F}(V_s)} = 0$ if and only if $\sum_{\alpha \in \mathfrak{F}(V_s)} \tilde{f}^{(\alpha)} = 0$ in $D(V_s/\tilde{V})$. From these remarks it follows that

$$M' = \left(\bigoplus_{\substack{V_i \subset V \\ V_i \cap \tilde{V} \neq 0, V_i}} D(V_i)\right) \oplus \left(\bigoplus_{\substack{V_{i+m} \subset V \\ \tilde{V} \subset V_{i+m}}} D(V_{i+m}/\tilde{V}) \underset{W_F}{\otimes} Q_i(V_{i+m}, \tilde{V})\right)$$

and

$$M'' = \left(\bigoplus_{V_i \subset \tilde{V}} D(V_i)\right) \oplus \left(\bigoplus_{\substack{V_{i+m} \subset V \\ \tilde{V} \subset V_{i+m}}} D(V_{i+m}/\tilde{V})\right).$$

Since the subspaces $V_{i+m} \subset V$ such that $\tilde{V} \subset V_{i+m}$ are in 1–1 correspondence with i-dimensional linear subspaces of $V/\tilde{V}$ the required formula for M'' follows.

Next we observe that if V_i is a linear subspace in V such that $V_i \cap \tilde{V} \neq 0, V_i$ then

$$D(V_i) = D(V_i/V_i \cap \tilde{V}) \underset{W_F}{\otimes} P_{W_F}(V_i, V_i \cap \tilde{V}) = D(V_i + \tilde{V}/\tilde{V}) \underset{W_F}{\otimes} P_{W_F}(V_i, V_i \cap \tilde{V})$$

(cf. 3.10). We put $V_s = V_i + \tilde{V}$, and collect together terms corresponding to the same V_s. The required formula for M' follows.

COROLLARY. *Let* Ω *be an algebraically closed field of characteristic zero, and let* $W_F \to \Omega$ *be a ring homomorphism. Then the* GL(V)*-modules* $\bigoplus_{V_i \subset V} D(V_i) \underset{W_F}{\otimes} \Omega$ *are irreducible for all* $1 \leq i \leq n$ *except when we have simultaneously* $q = 2$, $n \geq 2$, $i = 1$ *in which case* $\bigoplus_{V_1 \subset V} D(V_1) \underset{W_F}{\otimes} \Omega$ *has two irreducible components (one of which is the unit representation).*

This is an easy consequence of the proposition.

5.3 We shall now prove that, given $\tilde{V} \subset V$ $(0 < \dim \tilde{V} = m < \dim V = n)$ there exists a natural exact sequence of W_F-modules

$$(12) \qquad 0 \to (D(V))' \to \Big(\bigoplus_{V_{n-1} \subset V} D(V_{n-1})\Big)' \to \cdots \to \Big(\bigoplus_{V_1 \subset V} D(V_1)\Big)' \to 0$$

(the symbol M' has been defined in 5.2).

In fact, according to Proposition 5.2 it is sufficient to construct natural exact sequences

$$(13) \quad 0 \to P_{W_F}(V_s, \tilde{V}) \to Q_{s-1}(V_s, \tilde{V}) \to \cdots \to Q_{s-m+1}(V_s, \tilde{V}) \to Q_{s-m}(V_s, \tilde{V}) \to 0$$

for all $V_s \subset V$ such that $\tilde{V} \subset V_s$ and $m < s$. ((12) follows from (13) by tensoring with $D(V_s/\tilde{V})$ and taking direct sums over all V_s.) Now the natural exact sequence (13) is provided by 1.13(e) and our assertion follows. It is obvious that the exact sequence (12) is compatible with the action of $G_{\tilde{V}}^{V}$. It follows that

$$(14)\quad (-1)^{n-1}(D(V))'+(-1)^{n-2}\Big(\bigoplus_{V_{n-1}\subset V} D(V_{n-1})\Big)'+\cdots+(-1)^{0}\Big(\bigoplus_{V_1\subset V} D(V_1)\Big)'=0$$

in the Grothendieck group $R_{W_F}(G_{\tilde{V}}^{V})$.

Let now G be a finite group and assume that G is acting (linearly) in the vector space V. We shall show that the correspondence $V \to br(V)$ defines a group homomorphism $br: R_F(G) \to R_{W_F}(G)$. We have to show that for any exact sequence $0 \to \tilde{V} \to V \to V/\tilde{V} \to 0$ of F-vector spaces with G-action we have $br(V) = br(\tilde{V}) + br(V/\tilde{V})$ in $R_{W_F}(G)$. This follows from (14) and Proposition 5.2(ii).

We can state the following

THEOREM. *Let G be a finite group and let F be a finite field. Let V be a F-vector space on which G acts. Then there exists a canonical sequence of free W_F-modules with G action* $V^{(n)}, V^{(n-1)}, \ldots, V^{(1)} (n = \dim_F V)$ *and a canonical exact sequence of G-modules.*

$$0 \to V^{(n)} \otimes_{W_F} F \to V^{(n-1)} \otimes_{W_F} F \to \cdots \to V^{(1)} \otimes_{W_F} F \to V \to 0 .$$

Moreover the correspondence

$$V \to (-1)^{n-1} V^{(n)} + (-1)^{n-2} V^{(n-1)} + \cdots + (-1)^{0} V^{(1)}$$

defines a group homomorphism $br: R_F(G) \to R_{W_F}(G)$. *We have* $d \circ br =$ *identity map of* $R_F(G)$, *where* $d: R_{W_F}(G) \to R_F(G)$ *is the decomposition homomorphism induced by the canonical projection* $W_F \to F$.

5.4 We shall now determine the character of $br(V)$ on an arbitrary element of G (where V is an F-vector space with a G-action). It is clear that

$$\mathrm{Tr}_{W_F}(g|\mathrm{br}(V)) = \tilde{\mathrm{Tr}}(g|V)$$

whenever $g \in G$ has order prime to p. On the other hand we shall prove that

$$\mathrm{Tr}_{W_F}(g' \cdot g''|\mathrm{br}(V)) = \mathrm{Tr}_{W_F}(g'|\mathrm{br}(V))$$

whenever $g', g'' \in G$ are such that g' has order prime to p, g'' has order a power of p and $g' \cdot g'' = g'' \cdot g'$. We restrict the action of G on V to the subgroup G' of G generated by g' and g''; note that G' is a cyclic group. We can find a G'-invariant flag $V_{i_1} \subset V_{i_2} \subset \cdots \subset V_{i_k}$ of linear subspaces of V such that g'' is the identity on $V_{i_1}, V_{i_2}/V_{i_1}, \cdots, V/V_{i_k}$. We have $V = V_{i_1} + V_{i_2}/V_{i_1} + \cdots + V/V_{i_k}$ in $R_F(G')$. Since br is a group homomorphism (cf. 5.3), we have

$$\mathrm{br}(V) = \mathrm{br}(V_{i_1}) + \mathrm{br}(V_{i_2}/V_{i_1}) + \cdots + \mathrm{br}(V/V_{i_k}) \text{ in } R_{W_F}(G')$$

hence

$$\begin{aligned}\mathrm{Tr}_{W_F}(g'g''|\mathrm{br}(V)) &= \mathrm{Tr}_{W_F}(g'g''|\mathrm{br}(V_{i_1})) + \mathrm{Tr}_{W_F}(g'g''|\mathrm{br}(V_{i_2}/V_{i_1})) \\ &\quad + \cdots + \mathrm{Tr}_{W_F}(g'g''|\mathrm{br}(V/V_{i_k})) \\ &= \mathrm{Tr}_{W_F}(g'|\mathrm{br}(V_{i_1})) + \mathrm{Tr}_{W_F}(g'|\mathrm{br}(V_{i_2}/V_{i_1})) + \cdots + \mathrm{Tr}(g'|\mathrm{br}(V/V_{i_k})) \\ &= \mathrm{Tr}_{W_F}(g'|\mathrm{br}(V)).\end{aligned}$$

This shows that the function $g \to \mathrm{Tr}_{W_F}(g|\mathrm{br}(V))$ is precisely the Brauer lifting of the modular character of V as a G-module (see [5], [9]).

The fact that this is a virtual character of G was first pointed out by Green ([5], Theorem 1) and proved using Brauer's characterization of characters in terms of elementary subgroups [1]. (This has been called the Brauer lifting by Quillen [9], who showed how it can be applied to homotopy theory.)

5.5 In this section we show how Theorem 5.3 implies the following

THEOREM (Swan [17], Theorem 3). *Let* G *be a finite group and let* F *be a finite field. Then the natural homomorphism* $R_{W_F}(G) \to R_{Q_F}(G)$ *induced by the inclusion* $W_F \to Q_F$ *is an isomorphism.*

Note that Swan proves a more general result, where (W_F, Q_F) is replaced by (A,K) A any semilocal Dedekind ring, and K its field of fractions, but this is not needed here.

Proof. We define a map $R_{Q_F}(G) \to R_{W_F}(G)$ as follows. Let M be a Q_F-vector space with a G-action. We choose a G-invariant W_F-lattice L in M and we send $M \in R_{Q_F}(G)$ to $L \in R_{W_F}(G)$. We have to check that this is well defined, i.e., independent of the choice of L. It is sufficient to prove that given two G-invariant W_F-lattices L, L' in M such that $pL \subset L' \subset L$, we have $L = L'$ in $R_{W_F}(G)$. Using the isomorphism $R_{W_F}(G) \to R'_{W_F}(G)$ (see 5.1) we see that it is enough to prove that $L = L'$ in $R'_{W_F}(G)$. We have an exact sequence $0 \to L' \to L \to L/L' \to 0$. It is then enough to prove that $L/L' = 0$ in $R'_{W_F}(G)$. Since $pL/L' = 0$, it follows that L/L' is an F-vector space with a G-action. Using Theorem 5.3 we can find a sequence $V^{(n)}, V^{(n-1)}, \dots, V^{(1)}$ of free W_F-modules with G-action and an exact sequence of F-vector spaces compatible with the G-action:

$$0 \to V^{(n)} \underset{W_F}{\otimes} F \to V^{(n-1)} \underset{W_F}{\otimes} F \to \dots \to V^{(1)} \underset{W_F}{\otimes} F \to L/L' \to 0 .$$

In order to prove that $L/L' = 0$ in $R'_{W_F}(G)$ it is then enough to prove that $V^{(i)} \underset{W_F}{\otimes} F = 0$ in $R'_{W_F}(G)$, for all i, $1 \leq i \leq n$.

We have an exact sequence $0 \to V^{(i)} \xrightarrow{p} V^{(i)} \to V^{(i)} \underset{W_F}{\otimes} F \to 0$ hence $V^{(i)} \underset{W_F}{\otimes} F = V^{(i)} - V^{(i)} = 0$ in $R'_{W_F}(G)$. It follows that $L = L'$ in $R_{W_F}(G)$ hence our map $R_{Q_F}(G) \to R_{W_F}(G)$ is well defined. It is easy to check that this is just the inverse of $R_{W_F}(G) \to R_{Q_F}(G)$ and the theorem is proved.

COROLLARY 1. *Let* $L, L' \in R_{W_F}(G)$ *be such that* $Tr_{W_F}(g|L) = Tr_{W_F}(g|L')$, *for all* $g \in G$. *Then* $L = L'$ in $R_{W_F}(G)$.

COROLLARY 2. *The map* $br : R_F(G) \to R_{W_F}(G)$ *is a ring homomorphism.*

In fact, we must check that $br(V \otimes V') = br(V)\, br(V')$ in $R_{W_F}(G)$ for any F-vector spaces V, V' with G-action. According to Corollary 1, it is sufficient to check the equality of the corresponding traces, which is obvious.

5.6 Let V be a vector space of dimension $n \geq 1$ over F. It is well known that the Steinberg module $St_F(V)$ (see 1.13) is a projective indecomposable module for the group algebra $F[GL(V)]$.

THEOREM. *The* $F[GL(V)]$*-module* $V \underset{F}{\otimes} St_F(V)$ *is projective and indecomposable, provided* $q \neq 2$.

Proof. We can clearly assume that $n \geq 2$. The fact that $V \underset{F}{\otimes} St_F(V)$ is projective follows from the fact that $St_F(V)$ is projective. It follows that there exists a unique projective $W_F[GL(V)]$-module $\tilde{M}$ such that $\tilde{M} \underset{W_F}{\otimes} F = V \otimes St_F(V)$. $\tilde{M}$ has the following character:

$$Tr_{W_F}(t|\tilde{M}) = \tilde{T}r(t|V) \cdot Tr_{W_F}(t|St_{W_F}(V)),\quad t \in GL(V)\,.$$

We have

$$(15)\ \tilde{M} = D(V) \oplus \left(\bigoplus_{V_1 \subset V} D(V/V_1) \underset{W_F}{\otimes} St_{W_F}(V_1)\right) \oplus \cdots \oplus \left(\bigoplus_{V_{n-1} \subset V} D(V/V_{n-1}) \underset{W_F}{\otimes} St_{W_F}(V_{n-1})\right)$$

as $Q_F[GL(V)]$-modules. This can be checked by a character computation (the character of the Steinberg module is known, see [15]). For example the equality of the dimensions of the two sides of (15) is the identity

$$n \cdot q^{1+2+\cdots+(n-1)} = \sum_{i=0}^{n-1} q^{1+2+\cdots+(i-1)} \frac{(q^{i+1}-1)(q^{i+2}-1)\cdots(q^n-1)}{(q^{n-i}-1)} .$$

Replacing q by q^{-1} and multiplying with $q^{1+2+\cdots+(n-1)}$ this becomes

$$n = \sum_{i=0}^{n-1} (-1)^{n+i-1} \frac{(q^{i+1}-1)(q^{i+2}-1)\cdots(q^n-1)}{(q^{n-i}-1)}$$

which is precisely the identity expressing the vanishing of the alternating sum of dimensions of the terms in the exact sequence 1.14(c). Assume now that $V \underset{F}{\otimes} St_F(V)$ is not indecomposable. Then $\tilde{M}$ is not indecomposable. Since the GL(V)-modules $\bigoplus_{V_i \subset V} D(V/V_i) \underset{W_F}{\otimes} St(V_i)$ $(0 \leq i \leq n-1)$ are irreducible when the scalars are extended to any field of characteristic zero $(q \neq 2)$ we must have then the identity

$$(16) \quad mq^{1+2+\cdots+(n-1)} = \sum_{i=0}^{n-1} a_i\, q^{1+2+\cdots+(i-1)} \frac{(q^{i+1}-1)(q^{i+2}-1)\cdots(q^n-1)}{(q^{n-i}-1)}$$

where $0 < m < n$ and $a_i (0 \leq i \leq n-1)$ are numbers equal to 0 or 1, and such that $a_0 = 0$. (In fact, we can write $\tilde{M} = \tilde{M}_1 \oplus \tilde{M}_2$ with $\tilde{M}_1$ projective, $\tilde{M}_1 \neq 0, \tilde{M}$ and such that $D(V)$ occurs in $\tilde{M}_2$. It is well known that the dimension of $\tilde{M}_1$ must be of the form $m \cdot q^{1+2+\cdots+(n-1)}$). The identity (16) is impossible since its two sides have distinct p-adic valuations. The theorem is proved.

Remark. The case $n = 2$ of the theorem is due to V. Jeyakumar [7].

5.7 According to 4.10 the GL(V)-module $\bigoplus_{V_i \subset V} D(V_i)$ considered in 5.2 can be defined naturally over the discrete valuation ring $\mathcal{O}_{F,i} \subset K_{F,i}$ $(1 \leq i \leq n)$.

Let K_F be the subfield of Q_F generated by $\tilde{x}$, $x \in F^*$ and by $\lambda(F^n)$ for all $n \geq 2$. We have $K_F = \varinjlim_n K_{F,n}$. Let $\mathcal{O}_F = K_F \cap W_F$. Then $\mathcal{O}_F = \varinjlim_n \mathcal{O}_{F,n}$ and $\mathcal{O}_F$ is a discrete valuation ring in K_F with a unique prime ideal $(= p \cdot \mathcal{O}_F)$ with residue field F. Let $D(V)_{\mathcal{O}_F} = D(V)_{\mathcal{O}_{F,n}} \bigotimes_{\mathcal{O}_{F,n}} \mathcal{O}_F$. The formal alternating sum

$$\begin{aligned} br_{\mathcal{O}_F}(V) &= (-1)^{n-1} D(V)_{\mathcal{O}_F} \\ &+ (-1)^{n-2} \bigoplus_{V_{n-1} \subset V} D(V_{n-1})_{\mathcal{O}_F} + \cdots + (-1)^0 \bigoplus_{V_1 \subset V} D(V_1)_{\mathcal{O}_F} \end{aligned}$$

can then be regarded as an element in $R_{\mathcal{O}_F}(GL(V))$ (cf. 5.1). We thus get a natural Brauer-lifting homomorphism $R_F(G) \to R_{\mathcal{O}_F}(G)$ defined for all finite groups G. It is easy to check that all the proofs given in the case of the Witt ring W_F remain valid when W_F is replaced by $\mathcal{O}_F$. There is a natural commutative diagram

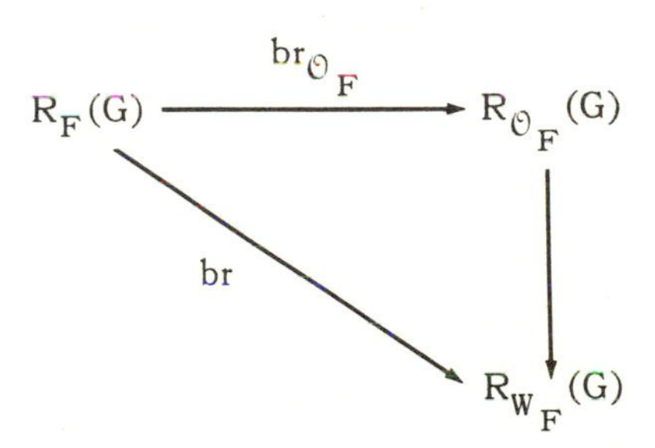

which shows that the map $br_{\mathcal{O}_F}$ is a refinement of br.

5.8 Let k be an integer. Consider the class function $\chi_{V,k}: GL(V) \to K_F$ defined as follows:

$$\chi_{V,k}(t) = (-1)^{n+j}\, \tilde{T}r(t^k | V_{an})(q^m-1)(q^{2m}-1)\cdots(q^{(j-1)m}-1) \tag{17}$$

where $t \in GL(V)$ is isotypic, $V_{an} \subset V$ is a t-invariant subspace of V such that $t|V_{an}$ is anisotropic and j is the number of terms in a decomposition of V into non-zero, indecomposable, t-invariant subspaces. On all other elements $t \in GL(V)$, define $\chi_{V,k}(t) = 0$.

It is clear that $\chi_{V,1}(t) = Tr_{W_F}(t|D(V))$. We shall prove now that for any $k \in \mathbf{Z}$, $\chi_{V,k}$ is the character of a virtual representation of $GL(V)$. For any i, $1 \leq i \leq n$ define a class function $\chi_{V,k}^{(i)} : GL(V) \to K_F$ by the formula

$$\chi_{V,k}^{(i)}(t) = \sum_{\substack{V_i \subset V \\ t(V_i) = V_i}} \chi_{V_i,k}(t|V_i) .$$

Then

$$\chi_{V,k}^{(n)} = \chi_{V,k} \quad \text{and} \quad \chi_{V,k}^{(i)} = \mathrm{Ind}_{G_{V_i}^V}^{GL(V)}(\rho_{V,k}) \; (1 \leq i \leq n-1)$$

where $\rho_{V_i,k} : G_{V_i}^V \to K_F$ is the class function defined by

$$\rho_{V_i,k}(t) = \chi_{V_i,k}(t|V_i), \quad t \in G_{V_i}^V .$$

We can assume by induction that $n \geq 2$ and that $\chi_{V',k}$ is the character of some virtual representation of $GL(V')$ for all F-vector spaces V' of dimension $< n$. Hence we can assume that $\chi_{V,k}^{(i)}$ is the character of some virtual representation of $GL(V)$ for all i, $1 \leq i \leq n-1$. We have the formula

$$(18) \quad (-1)^{n-1} \chi_{V,k}^{(n)}(t) + (-1)^{n-2} \chi_{V,k}^{(n-1)}(t) + \cdots + (-1)^0 \chi_{V,k}^{(1)}(t) = \tilde{T}r(t^k|V)$$

for all $t \in GL(V)$. Note that in case $k = 1$, (18) follows from 5.4. Assuming that (18) has been proved, we see that $\chi_{V,k}$ is a character of a virtual representation of $GL(V)$ if and only if $t \to \tilde{T}r_{W_F}(t^k|V)$ is. We have clearly $\tilde{T}r_{W_F}(t^k|V) = br(\psi^k V)$ where $\psi^k V \in R_F(GL(V))$ is obtained by applying to V the Adams operation ψ^k (see [9]). It follows then by induction that $\chi_{V,k}$ is the character of a virtual representation of $GL(V)$. We now prove (18). This is equivalent to the identity:

$$\sum_{\substack{V' \subset V \\ tV' = V' \\ V' \neq 0}} (-1)^{\dim V'+1} \chi_{V',k}(t|V') = \tilde{\mathrm{Tr}}(t^k|V) . \tag{19}$$

Write $V = V_{(1)} \oplus V_{(2)} \oplus \cdots \oplus V_{(h)}$ where $V_{(1)}, V_{(2)}, \cdots, V_{(h)}$ are the maximal t-invariant, isotypic subspaces of V. Let $\tau_k(t|V)$ be the left hand side of (19). Note that in the sum defining $\tau_k(t|V)$ the only terms which can be non-zero correspond to subspaces V' such that $t|V'$ is isotypic. Since any such V' is contained in one of the subspaces $V_{(1)}, V_{(2)}, \cdots V_{(h)}$ it follows that

$$\tau_k(t|V) = \tau_k(t|V_{(1)}) + \tau_k(t|V_{(2)}) + \cdots + \tau_k(t|V_{(h)}) .$$

It is obvious that this relation remains true if we replace $\tau_k(t|\cdot)$ by $\tilde{\mathrm{Tr}}(t^k|\cdot)$. In this way we see that it is sufficient to prove (19) in case $t|V$ is isotypic. In this case, let V_{an} be some non-zero t-invariant, subspace of V such that $t|V_{an}$ is anisotropic. We have

$$\tau_k(t|V) = \tilde{\mathrm{Tr}}(t^k|V_{an}) \cdot P_t(q)$$

where $P_t(q)$ is a polynomial in q independent of k. Moreover $P_t(q) = P_{t'}(q)$ if $t, t' \in GL(V)$ can be written as $t = su = us$, $t' = s'u' = u's'$ (s, s' of order prime to p; u, u' of order a power of p) so that $u = u'$, $s \in$ center of centralizer of s', and $s' \in$ center of centralizer of s. In particular we have

$$\tau_1(t|V) = \tilde{\mathrm{Tr}}(t|V_{an}) P_t(q) .$$

But for $k = 1$, we know that $\tau_1(t|V) = \tilde{\mathrm{Tr}}(t|V)$. Since $t|V$ is isotypic we have $\tilde{\mathrm{Tr}}(t|V) = \frac{n}{m} \tilde{\mathrm{Tr}}(t|V_{an})$ where $m = \dim_F (V_{an})$. It follows that

$$\tilde{\mathrm{Tr}}(t|V_{an})(P_t(q) - \tfrac{n}{m}) = 0 .$$

Using the fact that $P_t(q)$ depends only weakly on t, we can assume that $\tilde{\mathrm{Tr}}(t|V_{an}) \neq 0$ hence $P_t(q) = \frac{n}{m}$. We now return to the case when k

is arbitrary. We have $\tau_k(t|V) = \tilde{T}r(t^k|V_{an}) \cdot \frac{n}{m} = \tilde{T}r(t^k|V)$ since $t|V$ is isotypic and (19) is proved. The class function $\chi_{V,k}$ is in fact the character of an irreducible GL(V)-module for most values of k. In fact, a calculation shows that

$$\frac{1}{|GL(V)|} \sum_{t \in GL(V)} \chi_{V,k}(t)\chi_{V,k}(t^{-1}) = \frac{1}{q^n-1} \sum_{x \in F'^*} (1 + \tilde{x}^{a(q-1)} + \cdots + \tilde{x}^{a(q^{n-1}-1)})$$

$$= 1 + N(k)$$

where F' is an extension field of degree n of F and N(k) is the number of indices i, $1 \leq i \leq n-1$ such that

$$k(q^i - 1) \equiv 0 (\mathrm{mod}\ q^n - 1)\ .$$

This shows that $\chi_{V,k}$ is the character of an irreducible GL(V)-module if and only if $N(k) = 0$, that is, if and only if

$$(20) \qquad k(q^i-1) \not\equiv 0(\mathrm{mod}\ q^n-1), \text{ for all } i,\ 1 \leq i \leq n-1\ .$$

We can state the following

THEOREM (Green [5]). *For any integer* k *satisfying* (20) *there exists an irreducible* GL(V)*-module over some algebraically closed field of characteristic zero* $\Omega \supset K_{F,n}$ *whose character is* $\chi_{V,k}$. *If* k, k' *satisfy* (20), *we have* $\chi_{V,k} = \chi_{V,k'}$ *if and only if* $k' \equiv q^i k\ (\mathrm{mod}\ q^n-1)$ *for some* i, $0 \leq i \leq n-1$.

5.9 In this section we shall prove that for any integer k satisfying (20), the irreducible GL(V)-module with character $\chi_{V,k}$ occurs with multiplicity one in the GL(V)-module $\dot{\Delta}_K(V)$ where $K = K_{F,n}$ (see 4.10). We can assume $n \geq 2$. In fact, let $\varepsilon = (E_0 \subset E_1 \subset \cdots \subset E_{n-1})$ be some flag in Y and let B_ε be its stabilizer in GL(V). For any subset $I \subset \{0,1,\cdots,n-1\}$ let $B_{\varepsilon,I}$ be the stabilizer in GL(V) of the incomplete flag

$(E_{i_1} \subset E_{i_2} \subset \cdots \subset E_{i_k})$ where $I = \{i_1, i_2, \cdots, i_k\}$. For example $B_{\varepsilon,\emptyset} = GL(V)$. Let χ_I be the character of the GL(V)-module induced by the unit representation of $B_{\varepsilon,I}$. From the homological definition of $\dot{\Delta}_k(V)$ (see 1.13) and from the vanishing theorem for the homology of S_{III} (see 1.10) it follows easily that the character of the GL(V)-module $\dot{\Delta}_\Omega(V)$ is equal to the alternating sum $\sum_I (-1)^{|I|+n} \chi_I$ where the sum is over all subsets $I \subset \{0, 1, \cdots, n-1\}$; $\Omega \supset K$ is some algebraically closed field.

Next we observe that the restriction of $\chi_{V,k}$ to $B_{\varepsilon,I}$ is independent of k for any $I \neq \emptyset$. In fact in this case any element $t \in B_{\varepsilon,I}$ must have some eigenvalue equal to 1. If t is isotypic it follows that t must be in fact unipotent and our statement follows from (17); if t is not isotypic then $\chi_{V,k}(t) = 0$ for all k, hence $\chi_{V,k}(t)$ is again independent of k.

It follows that the inner product $\langle \chi_{V,k}, \sum_I (-1)^{|I|+n} \chi_I \rangle_{GL(V)}$ is independent of k (note that $\chi_\emptyset$ is the character of the unit representation of GL(V)). We want to prove that this inner product is equal to 1. We know already that this is true for $k=1$. (D(V) occurs with multiplicity one in $\mathrm{Ind}_{B_\varepsilon}^{GL(v)}(1)$ (cf. 2.1 and 3.8) and is contained in the subspace $\dot{\Delta}_\Omega(V)$.) It follows that the character $\chi_{V,k}$ occurs with multiplicity one in $\dot{\Delta}_\Omega(V)$ (and also in $\mathrm{Ind}_{B_\varepsilon}^{GL(V)}(1)$). Since $\dot{\Delta}_\Omega(V) \cong \dot{\Delta}_K(V) \otimes \Omega$, and $\chi_{V,k}$ has values in K our assertion is proved.

We can state the following.

THEOREM. *For any integer* k *satisfying* (20) *there is a unique* K*-linear subspace* $D_k(V)_K$ *of* $\dot{\Delta}_K(V)$ *which is a* GL(V)*-submodule with character* $\chi_{V,k}$ *given by* (17) $(K = K_{F,n})$. *Moreover, the character* $\chi_{V,k}$ *occurs with multiplicity one in* $\mathrm{Ind}_{B_\varepsilon}^{GL(V)}(1)$.

COROLLARY. *The character* $\chi_{V,k}$(k *satisfying* (20)) *can be realized by a* GL(V)*-module defined over the subfield of* $K_{F,n}$ *generated by the values of* $\chi_{V,k}$.

This follows from the fact that $\chi_{V,k}$ occurs with multiplicity one in a GL(V)-module defined over Q.

5.10 It follows from (17) that the GL(V)-module $D_k(V)_K$ becomes isomorphic to $\Delta_K(E)$ by restriction to the subgroup Aff(E) of GL(V) (where E is an affine hyperplane in $V, E \not\ni 0$). In particular $D_k(V)_K$ restricted to any p-subgroup of GL(V) is independent of k. It follows that $D_k(V)_K$ *belongs to the discrete series* of GL(V). In fact, we must check that $D_k(V)_K$ satisfies the cusp conditions, but these involve sums over certain p-subgroups of GL(V) and so the general case follows from the case $k = 1$, which is already known (cf. 3.12).

5.11 Since $D_k(V)_K$ occurs with multiplicity one in $\dot{\Delta}_K(V)$, it must be left invariant by any GL(V)-endomorphism of $\dot{\Delta}_K(V)$. In particular, $D_k(V)_K$ is left invariant by T (see 3.1). Since $D_k(V)_K$ is (absolutely) irreducible, T must be just multiplication by a constant on $D_k(V)_K$. In fact a proof completely similar to the one of Theorem 4.8 shows that

$$(21) \qquad Tf = \left(\sum_{\substack{x \in F' \\ Tr_{F'/F} x = 1}} \tilde{x}^{-k} \right) f$$

for all $f \in D_k(V)_K$, where $[F' : F] = n$.

It is easy to see that we must also have

$$(22) \qquad f(\lambda E_0 \subset \lambda E_1 \subset \cdots \subset \lambda E_{n-1}) = \tilde{\lambda}^{-k} f(E_0 \subset E_1 \subset \cdots \subset E_{n-1})$$

for all $f \in D_k(V)_K$, $\lambda \in F^*$ and $(E_0 \subset E_1 \subset \cdots \subset E_{n-1}) \in Y$.

CONJECTURE. *For any integer* k *satisfying* (20), $D_k(V)_K$ *is the set of all* $f \in \dot{\Delta}_K(V)$ *satisfying* (21) *and* (22).

This is true for $k = 1$ by the definition of $D(V)_K$. More generally, suppose that k and q^n-1 are relatively prime. Consider the field automorphism of $K_{F',1}([F':F] = n)$ defined by $\tilde{x} \to \tilde{x}^k$ where $\tilde{x}$ is any (q^n-1)-th root of unity in $K_{F',1}$. This induces an automorphism γ of the subfield $K_{F,n}$ (cf. 4.11) and it is easy to see that $D(V)_K^{\gamma} = D_k(V)_K$ (see 4.12). It follows then from 4.12 that the conjecture is true for all integers k which are relatively prime to q^n-1.

5.12 *Final remarks*

It would be interesting to study the Brauer lifting of modular representations of finite groups other than the general linear group. Consider, for example the symplectic group $Sp_{2n}(F)$ over the finite field F with q elements. Let V be the $2n$-dimensional F-vector space in which $Sp_{2n}(F)$ acts preserving a symplectic form. Then V is a modular representation of $Sp_{2n}(F)$. Its Brauer lifting $br(V)$ must be an element in $R_{W_F}(Sp_{2n}(F))$. The question is: how can one write explicitly $br(V)$ as an alternating sum of absolutely irreducible $Sp_{2n}(F)$-modules? The answer can be described as follows (assume $q \neq 2, 3$). One can write

$$br(V) = M_1 - M_2 + \cdots + (-1)^{n-1}M_n + (-1)^n M_n' + (-1)^{n-1} M_{n-1}' + \cdots + (-1)^1 M_1'$$

where $M_i, M_i' (1 \leq i \leq n)$ are free W_F-modules with $Sp_{2n}(F)$-action such that $M_i \underset{W_F}{\otimes} Q_F$, $M_i' \underset{W_F}{\otimes} Q_F$ $(1 \leq i \leq n)$ are absolutely irreducible $Sp_{2n}(F)$-modules. Moreover, among the modules $M_1, \cdots, M_n, M_1', \cdots, M_n'$ there is precisely one in the discrete series of $Sp_{2n}(F)$, namely M_n'. The modules M_i, M_i' have the rank given by the formula:

$$\mathrm{rank}_{W_F} M_i = \frac{(q^{2n-2i+2}-1)(q^{2n-2i+4}-1)\cdots(q^{2n}-1)}{q^i-1}$$

$$\mathrm{rank}_{W_F} M_i' = \frac{(q^{2n-2i+2}-1)(q^{2n-2i+4}-1)\cdots(q^{2n}-1)}{q^i+1}$$

We have $M_i = \bigoplus_{\substack{V_i \subset V \\ V_i \text{ isotropic}}} D(V_i)$, $(1 \leq i \leq n)$. If we denote $M'_n = D_{sp}(V)$, then M'_i $(1 \leq i \leq n-1)$ is a submodule of $\bigoplus_{\substack{V_{n-i} \subset V \\ V_{n-i} \text{ isotropic}}} D_{sp}(V^{\perp}_{n-i}/V_{n-i})$ defined as an eigenspace of a certain explicit endomorphism. These statements will be proved elsewhere.

REFERENCES

[1]. Brauer, R.: A characterization of the characters of groups of finite order. Ann. of Math., 57 (1953), 357-377.

[2]. Folkman, J.: The homology groups of a lattice. J. Math. Mech. 15 (1966), 631-636.

[3]. Gelfand, I. M. and Graev, M. I.: Construction of irreducible representations of simple algebraic groups over a finite field. Soviet Mat. Dokl. 3 (1962), 1646-1649.

[4]. Gelfand, S. I.: Representation of the full linear group over a finite field (Russian). Mat. Sbornik, 83 (125), (1970), 15-41.

[5]. Green, J. A.: The characters of the finite general linear groups. Trans. A.M.S. 80 (1955), 402-447.

[6]. Harish-Chandra: Eisenstein series over finite fields, Functional Analysis and Related fields. Edited by F. E. Browder, Springer 1970.

[7]. Jeyakumar, V.: Principal indecomposable representations for the group SL(2, q). Thesis, University of Madras.

[8]. Lusztig, G.: On the discrete series representations of the general linear groups over a finite field. Bull. A.M.S. 79 (1973), 550-554.

[9]. Quillen, D.: The Adams conjecture. Topology 10 (1971), 67-80.

[10]. Serre, J. P.: Corps locaux. Hermann, Paris 1968.

[11]. Solomon, L.: On the affine group over a finite field, Representation theory of finite groups and related topics. Proc. Symp. Pure Math. A.M.S. 1971.

[12]. ———————: The affine group I. Bruhat decomposition. J. of Algebra, 20 (1972), 512-539.

[13]. Springer, T. A.: Characters of special groups, Seminar on algebraic groups and related finite groups by A. Borel et al., Lecture Notes in Mathematics 131, Springer 1970.

[14]. ————: Caracteres des groupes de Chevalley finis. Sem. Bourbaki No. 429, 1972/73.

[15]. Steinberg, R.: A geometric approach to the representations of the full linear group over a Galois-field. Trans. A.M.S. 71 (1951), 225-235.

[16]. ————: Regular elements of semisimple algebraic groups. Publ. Math. I.H.E.S. No. 25 (1965), 49-80.

[17]. Swan, R.: The Grothendieck group of a finite group. Topology, 2 (1963), 85-110.

INDEX

LIBRARY OF CONGRESS CATALOGING IN PUBLICATION DATA

Lusztig, George, 1946-
The discrete series of GL_n over a finite field.

(Annals of mathematics studies, no. 81)
Bibliography: p.
1. Representations of groups. 2. Linear algebraic groups. 3. Series. 4. Fields, Algebraic.

I. Title. II. Series.
QA71.L848 512'.2 74-11058
ISBN 0-691-08154-9

ANNALS OF MATHEMATICS STUDIES

Edited by Wu-chung Hsiang, John Milnor, and Elias M. Stein

81. The Discrete Series of GL_n over a Finite Field, *by* GEORGE LUSZTIG
80. Smoothings of Piecewise Linear Manifolds *by* M. W. HIRSCH *and* B. MAZUR
79. Discontinuous Groups and Riemann Surfaces, *edited by* LEON GREENBERG
78. Strong Rigidity is Locally Symmetric Spaces *by* G. D. MOSTOW
77. Stable and Random Motions in Dynamical Systems: With Special Emphasis on Celestial Mechanics, *by* JÜRGEN MOSER
76. Characteristic Classes, *by* JOHN W. MILNOR *and* JAMES D. STASHEFF.
75. The Neumann Problem for the Cauchy-Riemann Complex, *by* G. B. FOLLAND *and* J. J. KOHN
74. Lectures on *p*-Adic *L*-Functions *by* KENKICHI IWASAWA
73. Lie Equations *by* ANTONIO KUMPERA *and* DONALD SPENCER
72. Introduction to Algebraic *K*-Theory, *by* JOHN MILNOR
71. Normal Two-Dimensional Singularities, *by* HENRY B. LAUFER
70. Prospects in Mathematics, *by* F. HIRZEBRUCH, LARS HÖRMANDER, JOHN MILNOR, JEAN-PIERRE SERRE, *and* I. M. SINGER
69. Symposium on Infinite Dimensional Topology, *edited by* R. D. ANDERSON
68. On Group-Theoretic Decision Problems and Their Classification, by CHARLES F. MILLER, III
67. Profinite Groups, Arithmetic, and Geometry, *by* STEPHEN S. SHATZ
66. Advances in the Theory of Riemann Surfaces, *edited by* L. V. AHLFORS, L. BERS, H. M. FARKAS, R. C. GUNNING, I. KRA, *and* H. E. RAUCH
65. Lectures on Boundary Theory for Markov Chains, *by* KAI LAI CHUNG
64. The Equidistribution Theory of Holomorphic Curves, *by* HUNG-HSI WU
63. Topics in Harmonic Analysis Related to the Littlewood-Paley Theory, *by* ELIAS M. STEIN
62. Generalized Feynman Amplitudes, *by* E. R. SPEER
61. Singular Points of Complex Hypersurfaces, *by* JOHN MILNOR
60. Topology Seminar, Wisconsin, 1965, *edited by* R. H. BING *and* R. J. BEAN
59. Lectures on Curves on an Algebraic Surface, *by* DAVID MUMFORD
58. Continuous Model Theory, *by* C. C. CHANG *and* H. J. KEISLER
57. Seminar on the Atiyah-Singer Index Theorem, *by* R. S. PALAIS
56. Knot Groups, *by* L. P. NEUWIRTH
55. Degrees of Unsolvability, *by* G. E. SACKS
54. Elementary Differential Topology (Rev. edn., 1966), *by* J. R. MUNKRES
52. Advances in Game Theory, *edited by* M. DRESHER, L. SHAPLEY, *and* A. W. TUCKER
51. Morse Theory, *by* J. W. MILNOR
50. Cohomology Operations, *lectures by* N. E. STEENROD, *written and revised by* D.B.A. EPSTEIN
48. Lectures on Modular Forms, *by* R. C. GUNNING
47. Theory of Formal Systems, *by* R. SMULLYAN
46. Seminar on Transformation Groups, *by* A. BOREL *et al.*
44. Stationary Processes and Prediction Theory, *by* H. FURSTENBERG
43. Ramification Theoretic Methods in Algebraic Geometry, *by* S. ABHYANKAR

P